Analysis of the Air Force Logistics Enterprise

Evaluation of Global Repair Network Options for Supporting the C-130

Ben D. Van Roo, Manuel Carrillo, John G. Drew,

Thomas Lang, Amy L. Maletic, Hugh G. Massey,

James M. Masters, Ronald G. McGarvey,

Jerry M. Sollinger, Brent Thomas, Robert S. Tripp

Prepared for the United States Air Force

PROJECT AIR FORCE

The research described in this report was sponsored by the United States Air Force under Contract FA7014-06-C-0001. Further information may be obtained from the Strategic Planning Division, Directorate of Plans, Hq USAF.

Library of Congress Cataloging-in-Publication Data

Analysis of the Air Force logistics enterprise : evaluation of global repair network options for supporting the C-130 / Ben D. Van Roo ... [et al.].
     p. cm.
  Includes bibliographical references.
  ISBN 978-0-8330-4957-5 (pbk. : alk. paper)
  1. Hercules (Turboprop transports)—Maintenance and repair—Management—Evaluation. 2. United States. Air Force—Equipment—Maintenance and repair—Management—Evaluation. 3. Airplanes, Military—United States—Maintenance and repair—Management—Evaluation. I. Van Roo, Ben D.

   UG1242.T7A63 2011
   358.4'483—dc22

                                                                                      2010048964

The RAND Corporation is a nonprofit institution that helps improve policy and decisionmaking through research and analysis. RAND's publications do not necessarily reflect the opinions of its research clients and sponsors.

RAND® is a registered trademark.

Published 2011 by the RAND Corporation
1776 Main Street, P.O. Box 2138, Santa Monica, CA 90407-2138
1200 South Hayes Street, Arlington, VA 22202-5050
4570 Fifth Avenue, Suite 600, Pittsburgh, PA 15213-2665
RAND URL: http://www.rand.org/
To order RAND documents or to obtain additional information, contact
Distribution Services: Telephone: (310) 451-7002;
Fax: (310) 451-6915; Email: order@rand.org

Preface

Since the 1990s, the Air Force has continually engaged in deployed operations in Southwest Asia and other locations. Recent Office of the Secretary of Defense planning guidance directed the services to plan for high levels of engagement and deployed operations, although their nature, locations, durations, and intensity may be unknown. Recognizing that this new guidance might impose different demands on the logistics system, senior Air Force logistics leaders asked RAND Project AIR FORCE to undertake a logistics enterprise analysis. The objective was to identify and rethink the basic issues and the premises on which the Air Force plans, organizes, and operates its logistics enterprise.

The complex nature of this project led Project AIR FORCE researchers to approach the analysis in phases. Our previous analysis assessed the F-16 and KC-135 maintenance network.[1] This report evaluates the rebalancing of the C-130 maintenance network and focuses on determining the future logistics workload requirement and identifying alternatives for how the workload should be accomplished. We examine maintenance workloads of home-station units and forward operating locations (FOLs). We separated the workload into two categories: the mission-generation workload, which must be performed at an operating location; and network workload, which could be assigned to a supporting network location and may benefit from economies of scale in the maintenance process. We assessed the effects of a wide range of options, from a fully decentralized network, in which unit-level maintenance stays at each unit, to a fully centralized network, in which centralized repair facilities (CRFs) conduct some scheduled maintenance tasks, off-equipment component repair (both at home station and while deployed), and deferred nongrounding workloads from deployed environments. The objective of the CRF is to focus resources on inspecting and repairing the aircraft and components in a more efficient and effective manner than does a stand-alone operational unit.

In this report, we describe the costs and benefits of developing a CRF network, allocating workloads to it, and rebalancing the maintenance personnel within it. The focus of our research was on a network that would support only active-duty and Air Force Reserve

---

[1] See Ronald G. McGarvey, Manuel Carrillo, Douglas C. Cato, Jr., John G. Drew, Thomas Lang, Kristin F. Lynch, Amy L. Maletic, Hugh G. Massey, James M. Masters, Raymond A. Pyles, Ricardo Sanchez, Jerry M. Sollinger, Brent Thomas, Robert S. Tripp, and Ben D. Van Roo, *Analysis of the Air Force Logistics Enterprise: Evaluation of Global Repair Network Options for Supporting the F-16 and KC-135*, Santa Monica, Calif.: RAND Corporation, MG-872-AF, 2009.

Robert S., Tripp, Ronald G. McGarvey, Ben D. Van Roo, James M. Masters, and Jerry M. Sollinger, *A Repair Network Concept for Air Force Maintenance: Conclusions from Analysis of C-130, F-16, and KC-135 Fleets*, Santa Monica, Calif.: RAND Corporation, MG-919-AF, 2010, provides an executive summary covering the work described both in McGarvey et al. 2009 and in this report.

Command (AFRC) aircraft. We also present our findings for the Total Force network, which includes active-duty, AFRC, and the Air National Guard (ANG).[2]

This report also discusses an initial assessment of integrated maintenance concepts, such as high-velocity maintenance (HVM), that integrate wing-level and depot-level maintenance workloads.[3] The intention was to determine the order of magnitude of the potential benefits and savings of HVM to the Air Force.

The Deputy Chief of Staff for Logistics, Installations and Mission Support (AF/A4/7), along with the Vice Commander, Air Force Materiel Command (AFMC/CV), sponsored this research, which was carried out in the Resource Management Program of RAND Project AIR FORCE under two FY08 projects: "Global Materiel Management Strategy for the 21st Century Air Force," and "Managing Workload Allocations in the USAF Global Repair Enterprise."

This and other Project AIR FORCE logistics enterprise analyses have provided analytical support to the Air Force's Repair Network Integration (RNI) initiative, which seeks to provide integrated support to the warfighter through an enterprisewide repair capability managed by a single process owner. In 2009, Phase 1 of RNI examined core processes to manage and allocate enterprise workloads for a select group of weapon systems and subsystems. In 2010, Phase 2 of RNI began working with the entire propulsion enterprise. The objective of Phase 2 is to implement the processes, capabilities, and control of the network repair by 2015. This report should interest logistics and operational personnel throughout the Department of Defense (DoD) and those involved in logistics requirements determination.

RAND Project AIR FORCE

RAND Project AIR FORCE (PAF), a division of the RAND Corporation, is the U.S. Air Force's federally funded research and development center for studies and analysis. PAF provides the Air Force with independent analyses of policy alternatives affecting the development, employment, combat readiness, and support of current and future aerospace forces. Research is conducted in four programs: Force Modernization and Employment; Manpower, Personnel, and Training; Resource Management; and Strategy and Doctrine.

Additional information about PAF is available on our website:
http://www.rand.org/paf/

---

[2]   Congress imposed a number of restrictions on maintenance consolidations involving ANG units, including a requirement that the Secretary of Defense certify that a consolidation was in the national interest and that it would not harm recruiting and retention in the ANG (Duncan Hunter National Defense Authorization Act for Fiscal Year 2009, Section 324 of H.R. 5658, 110th Congress, 2008). Therefore, we indicate what the savings would be from consolidating selected maintenance activities for only AFRC and active-duty units while also showing what savings would accrue if the total force were factored into centralization.

[3]   HVM is a concept that Warner Robins Air Logistics Center (ALC) is implementing. It integrates wing- and depot-level workloads by incorporating workload into depot maintenance processes with the goal of increasing the "touch" labor productivity rates of the depot maintenance worker and increasing the flow rate of the programmed depot maintenance process.

Contents

Preface iii

Figures vii

Tables ix

Summary xi

Acknowledgments xvii

Abbreviations xxi

CHAPTER ONE

Introduction 1
Background 1
Research Motivation 1
    Guidance Changes 2
    Resource Reductions 3
Research Purpose, Objectives, and Approach 4
    What Will the Logistics Workload Be? 4
    How Should the Logistics Workload Be Accomplished? 4
    How Should These Questions Be Revisited Over Time? 5
    Answering the Questions 5
Focus of This Report 6
Organization of This Report 7

CHAPTER TWO

Projecting Logistics Workloads 9

CHAPTER THREE

Alternatives for Rebalancing C-130 Maintenance Resources 13
Determination of Maintenance Workload and Manpower Requirements 13
    Scheduled Maintenance 16
    Deferrable Maintenance 22
    Off-Equipment Component Repair 24
    Mission-Generation Maintenance 24
C-130 Component Repair Pipeline 29
C-130 Repair Network Design Options 30
    Total Manpower Requirements for the Mission-Generation and Network Facilities 34
CRF Networks to Support the Total Force 35
AFSOC Centralized ISO Inspection Facility 38

CHAPTER FOUR
Assessment of the Effects of Integrated Maintenance.. 41
High-Velocity Maintenance.. 41
Current Depot-Level Performance.. 42
Preliminary Analysis of High-Velocity Maintenance.. 43

CHAPTER FIVE
Conclusions.. 47

APPENDIXES
A. **Maintenance Manpower Authorizations**.. 49
B. **Analysis of ISO Inspections and HSCs Using REMIS**.. 55
C. **Modeling C-130 Maintenance with the Logistics Composite Model**.. 59
D. **Integer Linear Programming Model**.. 65

Bibliography.. 69

Figures

1.1.	Distribution of Standard and Specialty C-130 Aircraft	7
2.1.	Notional Flying Hour Requirements	10
3.1.	C-130 Distribution of Fly-to-Fly Times Organized by MAJCOM	17
3.2.	C-130 Economies of Scale in the ISO Inspection Process	18
3.3.	Manpower Utilization for the ISO Inspection Process	19
3.4.	ISO Inspection Flow Times as a Function of Size of CRF Operations	21
3.5.	Economies of Scale: AMC's RMF and RAND's CRF Concepts	23
3.6.	Flow Times of AMC's RMF Concept and RAND's CRF Concept	23
3.7.	FY 2008 C-130 Active-Duty and AFRC Worldwide Beddown	31
3.8.	Comparative Cost of C-130 Network Designs	33
3.9.	Total Cost Comparison of Current System and CRF Concept	34
3.10.	FY 2008 C-130 Active, AFRC, and ANG Worldwide Beddown	35
3.11.	Fly-to-Fly Comparison of Decentralized and Centralized ISO Inspections for AFSOC C-130s	39
3.12.	Comparison of LCOM-Estimated and Contractor Centralized ISO Inspection Facility	40
4.1.	Histogram of the Distribution of PDM Man-Hours, FY 2005–July 2008	43
4.2.	Comparison of CRF and HVM Concept for the Active-Duty and AFRC Network	45
C.1.	LCOM-Generated ISO Inspection Flow Times at an ISO Inspection–Only RMF	63

Tables

S.1.   Savings from CRF and HVM Processes.. xiv
3.1.   AMC RMF/FOL C-130 LCOM Workload Allocation..................................... 15
3.2.   RAND CRF C-130 LCOM Workload Allocation... 25
3.3.   Distribution of Mission-Generation Maintenance Work Centers for
        Standard C-130s.. 26
3.4.   Manpower Requirements for Standard C-130 Mission-Generation Operations.......... 28
3.5.   Manpower Requirements for Specialty C-130 Mission-Generation Operations.......... 28
3.6.   C-130 Active-Duty and AFRC CRF Network Options.................................. 32
3.7.   Manpower Requirements for the Total Force Mission-Generation Operations........... 36
3.8.   C-130 Total Force CRF Network Options... 37
4.1.   Warner Robins C-130 PDM Process Flow Data: FY 2007–July 2008.................... 43
4.2.   Comparison of CRF Concept and HVM Concept for the Active-Duty and
        AFRC Network.. 45
A.1.   C-130 Maintenance Work Centers.. 50
A.2.   Associate Unit Arrangements.. 51
A.3.   Standard C-130 Maintenance Personnel Authorization Totals 51
A.4.   Variant C-130 Maintenance Personnel Authorization Totals 51
A.5.   Manpower Requirements for Active-Duty and AFRC C-130 Network 53
A.6.   Manpower Requirements for Active-Duty and AFRC C-130 Network Supporting
        Split Operations.. 53
A.7.   Manpower Requirements for the Total Force C-130 Network.......................... 54
B.1.   C-130 ISO Inspection Fly-to-Fly Times, On-Aircraft Flow Times, and
        Maintenance Man-Hours, by MAJCOM 57
B.2.   C-130 HSC Fly-to-Fly Times, On-Aircraft Flow Times, and Maintenance
        Man-Hours, by MAJCOM... 57

Summary

Background and Purpose

Since the advent of the Expeditionary Air and Space Force concept in the 1990s, the Air Force has undergone numerous transformational processes and initiatives. Many of the initiatives have been local and the changes incremental. The independent effect of each incremental change is likely positive, but it is unknown whether the combined effects, viewed from a systems perspective, align with senior leaders' desired future direction for the Air Force. In 2007, senior Air Force logisticians asked RAND Project AIR FORCE to undertake a strategic reassessment of the Air Force's logistics enterprise to support the Air Force's efforts to realign the enterprise with the realities of the national security environment. A key part of this analysis was to identify alternatives for appropriately rebalancing logistics resources and capabilities between operating units and support network nodes across the total force.

The logistics enterprise analysis has four major objectives. The first is evaluating Department of Defense planning guidance to determine projected logistics system workloads. The second is structurally reviewing scheduled and unscheduled maintenance workloads that may be rebalanced between operating units and support networks. The third is strategically reevaluating the objectives and roles of contract as opposed to organic support in the logistics enterprise. The fourth is performing a top-down review of the management of the logistics transformation initiatives to ensure their integrated alignment with broader logistics objectives.

This report primarily focuses on the first two project objectives, in the context of the Air Force fleet of C-130s, including all Air Force standard C-130Es, C-130Hs, C-130Js, and all Air Force specialty variants including, among others, AC-130s, EC-130s, HC-130s, and MC-130s.[1] However, we also discuss an example of contractor support to the C-130 logistics enterprise. In addition, we discuss one transformation initiative, HVM, that could strongly influence the structuring of network-based maintenance workload between network nodes of differing capabilities.

We examined the C-130 fleet for several reasons. First, the C-130 is used extensively to support deployed missions. Earlier analyses of the F-16 and KC-135 fleets showed that the efficiency gains that could be achieved by consolidating workloads via a network of CRFs are significantly influenced by the maintenance strategy for deployed forces and the envisioned future deployment requirements for an aircraft fleet.[2] Second, the C-130 fleet is distinctive in

---

[1] McGarvey et al., 2009, addressed logistics support to the F-16 and the KC-135 and presented alternatives for the reallocation of maintenance workload between operating locations and a network of CRFs.

[2] McGarvey et al., 2009.

that it has several variants and supports an array of missions. We wanted to examine how the differences of the variants may affect centralizing the maintenance workload. Third, a portion of the C-130 fleet has already implemented the CRF concept. Air Force Special Operations Command (AFSOC) at Hurlburt Field centralized the isochronal (ISO) inspection process across several operating locations in the continental United States and a European operating location. AFSOC's motivation for centralization of ISO inspections was to free Air Force maintenance personnel from backshop work so they could be allocated to flight line sortie-generation work. The performance of the command's centralized ISO inspection facility provided strong empirical data to support our analysis. Moreover, AFSOC's use of a contractor to perform the centralized workload further develops the discussion of who could perform the maintenance at the CRFs.

A primary goal of this document and the document series is to develop a range of robust alternative policy solutions. The solutions will be used to inform the leadership of the Air Force of a range of options, rather than a single "best" or "optimal" answer. External factors beyond the scope of these analyses will influence the actual design of these networks. Therefore, the Air Force leadership could use the range of solutions provided in this document to weigh the costs and benefits of design alternatives and external factors when choosing the desired levels of capabilities and investment and the network design.

Results

This report presents a methodology for allocating maintenance workloads between operating units and a network of CRFs and examines the effects of centralizing specific maintenance workloads to achieve manpower economies of scale across the fleet while accounting for the manpower diseconomies associated with splitting other workloads between operating locations and CRFs, the costs of establishing CRFs, and the costs of transporting aircraft and components to and from CRFs. This analysis goes further than that of our other work because it also identifies the effects of maintenance consolidation on aircraft availability.[3]

For the purposes of illustration, we assumed that the desired capability is to support a steady-state deployment of 40 percent of the standard combat-direct support (CA) C-130 fleet, and 60 percent of the specialty CA and combat-coded (CC) fleet for an indefinite duration. For the F-16, our earlier analysis identified a requirement for a deployable CRF capability to support forward operations within an area of responsibility.[4] However, similar to the 2009 KC-135 analysis, which was based on current deployed maintenance practices for mobility aircraft and the existing Air Mobility Command (AMC) FOL–Regional Maintenance Facility (RMF) concept, this C-130 analysis assumes the CRF workload would be performed outside the deployed area of responsibility at a permanent location (which could be either inside or outside the continental United States). This analysis extends AMC's FOL-RMF concept by applying the FOL approach to home-station operations and by broadening the CRF workload to include ISO inspections, component repair, the refurbishment process workload, and some FOL nongrounding failure workload. Our analysis led to the following findings:

---

[3]   McGarvey, et al., 2009.

[4]   McGarvey, et al., 2009.

Rebalancing the workload with the introduction of CRFs could increase the capabilities and availability of Standard and Specialty C-130 aircraft while providing financial savings to the Air Force. First, we considered a CRF network that would support only the active-duty and AFRC network, with the ANG maintaining its current structure. We found that, if the Air Force implemented such a CRF concept and concluded that the current C-130 mission-generation maintenance operational capabilities were sufficient, the potential savings would be 2,500 personnel authorizations. The reduction in resources would generate an estimated $102 million annually by rebalancing workloads from operating locations to the network. Alternatively, the Air Force could elect to increase mission generation maintenance capabilities by authorizing a "split-operations" maintenance capability at CC/CA units.[5] Although units have been supporting split operations in recent experience, the additional maintenance manpower required to fully support both the deployed and home station responsibilities has not been authorized to these units. If the Air Force elected to apply the $102 million savings to split-operations requirements, the money could provide approximately 1,600 positions out of a 2,400-position split-operations requirement across all CC/CA units, without exceeding the current costs of the system.[6] To provide each active-duty AFRC CC/CA squadron with a split-operations capability (assuming the CRF concept is implemented), an additional 800 (2,400 − 1,600) authorizations need to be reassigned into C-130 maintenance, and the annual cost of C-130 maintenance would increase by approximately $26 million above the current level of expense.

Whether the Air Force captured the savings or reinvested in mission-generation maintenance, an expected 19 fewer aircraft would be in the ISO inspection process at any time. This translates to potentially making an additional 4.8 percent of the active-duty and AFRC total aircraft inventory (TAI) available to the Air Force for operational use.[7] This reduction is associated with a movement from one-shift ISO inspection operations at each unit to two- and three-shift operations at CRFs.

If we consider the Total Force, the CRF concept offers additional benefits and potential savings. Approximately two-thirds of all standard and specialty aircraft are possessed by active-duty and AFRC units (44 percent and 21 percent, respectively), and the remainder by the ANG. If the Air Force believed that the current mission-generation capabilities were adequate, consolidation across the total force could save 3,200 maintenance manpower authorizations, which would translate into a savings of $103 million annually. Alternatively, the resource savings from implementing the CRF concept could fund full-time split-operations maintenance positions. Providing every CC/CA squadron in the total force a split-operations capability would require transferring a total of 4,600 authorizations to the squadrons and increasing Air Force system costs by $120 million. Implementing the CRF concept would

---

[5] *Split operations* occur when only part of a squadron deploys and part remains at home station, thus splitting the unit. Mission generation maintenance personnel are required at both locations.

[6] The $102 million annual savings reflects a 2,500 position reduction that is a mix of full- and part-time authorizations, offset by increased expenditures for CRF construction and transportation. Our equation of this $102 million to 1,600 split-operations positions assumed that these positions were all active-duty maintainers. We also had to account for some additional facility and transportation costs. A mixture of active-duty and AFRC full- and part-time personnel could increase the number of split-operations authorizations that could be funded without exceeding current expenditures.

[7] Note that, while 19 fewer aircraft would be in the ISO process at any time, not all these aircraft would be expected to be available, since some of the 19 aircraft would likely be unavailable for other reasons, such as engine failures.

generate approximately 1,600 full-time split-operations maintenance positions at a funding level that matches the current system costs. However, if we expand our perspective beyond this single weapon system, application of the CRF concept more broadly might allow for savings generated from aircraft with lesser deployment burdens, such as fighters, to be invested in a split-operations capability for other weapon systems with a greater deployment burden. Of course, were the leadership to deem the rotational burdens associated with aircraft maintenance acceptable, it could then reallocate any or all of these manpower "savings" out of aircraft maintenance and into other needs within the Air Force. From an effectiveness perspective, an expected 35 fewer aircraft would be in the ISO inspection process at any point in time. This translates to an additional 5.9 percent of the total force TAI that would potentially be available for Air Force operational missions. The reduction of the number of aircraft in the ISO inspection process illustrates the effect of implementing two- and three-shift operations instead of the large number of one-shift ISO inspection operations in the ANG and AFRC.

A number of potential network configurations with alternative CRF locations have comparable total system costs and availability effects. Our analysis determined that many potential sets of CRF sites provide benefits comparable to those of the optimal solution. Air Force leadership can use the range of options discussed in this research to weigh the design alternatives, along with external factors not considered in this analysis, when choosing the desired levels of capabilities and investment and the network design.

Our initial assessment of the HVM concept shows that it can provide savings beyond those the CRF concept offers. Integrating some of the wing-level workload into the depot-level workload for the active-duty and AFRC units could save approximately 3,300 maintenance authorizations, with a financial savings of $132 million. This translates into a savings of approximately 800 authorizations beyond the CRF concept. An estimated 52 fewer aircraft would be in the programmed depot maintenance (PDM)–ISO inspection process on average, twice the aircraft of implementing the CRF concept alone. This translates to an additional 13 percent of the active-duty and AFRC TAI that would potentially be available to the Air Force. If the total force is considered, the savings would increase to an estimated 4,500 maintenance authorizations and a total cost savings of $176 million. An estimated 86 fewer aircraft would be in the PDM–ISO inspection process on average. This translates to 14.5 percent of the total force TAI that would potentially be available to the Air Force. Table S.1 shows the savings from CRFs and HVM.

Finally, we emphasize again that a range of alternatives lies between the analysis endpoints that trade off additional split-operations capabilities with total cost savings. The Air Force

Table S.1
Savings from CRF and HVM Processes

Process	Cohort	Authorizations (number)	Annual Savings ($M)
CRF	Active-duty and AFRC	2,500	102
	Total force	3,200	103
HVM	Active-duty and AFRC	3,300	132
	Total force	4,500	176

could broaden its view of rebalancing resources across several mission design series (MDSs) to best meet the requirements of the future security environments. Similarly, rebalancing options should also consider the reprogramming of resources among maintenance and other career fields, based on projections of relative levels of future demand. Review and assessment of Office of the Secretary of Defense guidance, such as the Steady State Security Posture, could help the Air Force make such discriminations among aircraft and across operational capabilities.

Acknowledgments

Many people inside and outside the Air Force provided valuable assistance and support to our work. We list them here with their rank and organization as of the time of the research. We thank Lt Gen Kevin Sullivan, Headquarters U.S. Air Force, Deputy Chief of Staff, Logistics, Installations and Missions Support (AF/A4/7), Lt Gen Terry Gabreski (AFMC/CV), and Michael Aimone (AF/A4/7), who sponsored this research and continued to support it through all phases of the project.

On the Air Staff, we thank Maj Gen Robert McMahon, Director of Logistics, Deputy Chief of Staff for Logistics, Installations and Mission Support; Maj Gen Gary McCoy, Director of Logistics Readiness, Deputy Chief of Staff for Logistics, Installations and Mission Support, and then Commander Air Force Global Logistics Support Center; and Grover Dunn, Headquarters U.S. Air Force, Director of Transformation, Deputy Chief of Staff for Logistics, Installations and Mission Support, along with their staffs. Their comments and insights have sharpened this work and its presentation. We are grateful to our project officer, Lt Col Dave Koch, for his support and contributions, as well as to Brent Phillips, our other primary contact in the Air Staff. In addition, we would like to thank the former project officer, Lt Col Cheryl Minto, for her many contributions to this effort.

At Air Mobility Command, we thank Brig Gen Kenneth Merchant, Director of Logistics, Capt Jerrymar Copeland, and their staffs for providing support to our analysis. At AFSOC, we would like to thank Brig Gen (Select) John Cooper, Col Robert Burnett, Col Michael Vidal, Lt Col Paul Wheeless, Maj Mark Ford, Capt Mark Gray, and CMSgt Gerald Lautenslager. We were also fortunate to receive tremendous assistance from Maj Gen Thomas Owen, Director of Logistics and Sustainment, AFMC, and his staff throughout the entire research process.

At the Air National Guard Bureau, we thank Col Rich Howard, Director of Logistics, and his staff, including Col Steph Dowling, Col Tom Redford, Col Dave Whipple, Col Tom Murgatroyd, Col Chuck Melton, Col Wayne Shanks, and Col Gary Nolan, for sharing their comments and insights.

At AFRC, we received valuable feedback from Brig Gen Elizabeth Grote, Director of Logistics, Col Faylene Wright, and their staffs.

We visited many flying units and other organizations during the course of this study and received valuable information and feedback from all. Specifically, we would like to thank the following:

- 1st Special Operations Maintenance Operations Squadron, Air Force Special Operations Command, Hurlburt Field: Maj Raquel Wasilausky, Commander
- 182nd Maintenance Group: Lt Col Bart Welker, Commander

- 314th Maintenance Squadron, Little Rock AFB: Lt Col Steven Weld, Deputy Commander; Maj Elizabeth Clay, Maintenance Supervision; Capt James Delph, Maintenance Flight; MSgt Earnest Heflin, Accessories Maintenance Flight
- 403rd Maintenance Group, Keesler Air Force Base (AFB): Col Anthony Baity, Commander, and Lt Col Constance Von Hoffman, Deputy Commander
- 436th Maintenance Squadron, Dover AFB: Col Dennis Daley, Commander
- 436th Maintenance Group, Dover AFB: Maj David Glass, Commander
- 463rd Aircraft Maintenance Squadron: Maj Andrew A. Burke, Materials Laboratory
- 758th Maintenance Group:Lt Col Kenneth D. Honaker, Commander
- 911th Maintenance Group, Pittsburgh: CMSgt Terrance Keblish
- 919th Maintenance Group, Duke Field: Col James Brock, Commander; Lt Col David Booher, Deputy Commander; Capt Carmel Weed, Aircraft Maintenance Operations Officer; MSgt Claude Stuteville, Maintenance Flight.

Warner Robins ALC provided exceptional support under the guidance of its commander, Maj Gen Polly Peyer, as well as David Nakayama and other personnel, including Dale Foster, Randy Ivey, Jerry Mobley, Denise Bryant, and Samantha Knapp.

We also visited a number of deployed units, where we learned a great deal about the special challenges associated with aircraft maintenance in the deployed environment. We are grateful to the following:

- Lt Gen Gary North, Commander U.S. Air Forces Central, who sponsored the visit, and Col Peter Hunt, Chief of Staff, U.S. Air Forces Central, Shaw AFB, S.C., who helped to coordinate our travel to the deployed units
- Col Perry Oaks and Lt Col James Bruns, 332nd Electronics Maintenance Group, and Lt Col Robert Lepper and Capt Timothy Casey, 379th Electronics Maintenance Group, for allowing us to visit their maintenance operations
- Col David Carrell, Deputy Director of Logistics, U.S. Air Forces Central, and his entire staff, including Lt Col David Yockey, MSgt Thomas Colvin, Central Logistics Director Superintendent of Maintenance Manpower, U.S. Air Forces Central, for the outstanding assistance that they provided, both in advance of our visit and during our travel within Southwest Asia
- MSgt Ryan Fondulis, who accompanied the RAND team during our site visits.

We benefited from conversations with Col Brent Baker, Provisional Commander, Air Force Global Logistics Support Center, at Scott AFB. Rogelio Hudson at Mobility Air Forces Logistics Support Center, Scott AFB, shared with us his expert knowledge of Air Force Data Systems and pointed us to other points of contact. Arthur Eggleton, at Supply Chain Management Branch, AFMC, helped us get the weight and cube data associated with the shipping of aircraft components. John Cilento in the Flight Management Branch of Air Combat Command, provided insights into the relationship between programmed training flying hours and deployed flying hours.

We received invaluable support from Dick Enz and the Resources Management Information System (REMIS) office located at Wright-Patterson AFB. They provided extensive amounts of data in a timely manner and answered several questions along the way.

Our Logistics Composite Model analyses were greatly aided by the support we received from Shenita Clay, Logistics Composite Modeling Flight, 3rd Manpower Requirements Squadron, at Scott AFB.

At RAND, Candice Riley provided exceptional technical programming support to our analysis of REMIS maintenance data. We benefited greatly during our project from the inputs, comments and constructive criticism of many RAND colleagues, including (in alphabetical order) Laura Baldwin, Natalie Crawford, Greg Hildebrandt, Patrick Mills, Nancy Moore, Ricardo Sanchez, Don Snyder, and Donald Stevens. We would also like to thank 2007 RAND Air Force Fellows Col Doug Cato and Lt Col Lee Flint for their support and expertise. Finally, we thank Megan McKeever for her assistance throughout the production of this report.

As always, the analysis and conclusions are the responsibility of the authors.

Abbreviations

AB	air base
ACC	Air Combat Command
AETC	Air Education and Training Command
AFB	Air Force base
AFCENT	U.S. Air Forces Central
AFMC	Air Force Materiel Command
AFPE	Air Force program element
AFRC	Air Force Reserve Command
AFSOC	Air Force Special Operations Command
AGE	aircraft ground equipment
ALC	air logistics center
AMC	Air Mobility Command
AMXS	aircraft maintenance squadron
ANG	Air National Guard
APG	airplane general mechanic
AR	Air Reserve
BNRTS	base not reparable this station
BRAC	Base Realignment and Closure
BSP	baseline security posture
CA	combat-direct support
CC	combat-coded
CMS	component maintenance squadron
CONUS	continental United States
CRF	centralized repair facility

DoD	Department of Defense
ECM	electronic countermeasures
E&E	electrical and environmental
EMS	equipment maintenance squadron
FAC	functional account code
FOL	forward operating location
FY	fiscal year
FYDP	Future Years Defense Program
GAC	guidance and control
HSC	home-station check
HVM	high-velocity maintenance
IAP	international airport
ISO	isochronal
JEIM	jet engine intermediate maintenance
LCOM	Logistics Composite Model
MAF	man-hour availability factor
MAJCOM	major command
MCO	major combat operation
MDS	mission design series
MMH	maintenance man-hours
MPES	Manpower Programming and Execution System
MOS	maintenance operations squadron
MX	maintenance
MXG	maintenance group
MXS	maintenance squadron
NDI	nondestructive inspection
NGB	National Guard Bureau
NGF	nongrounding failure
NIIN	national item identification number
OPTEMPO	operating tempo

ORGT	organizational title
OSD	Office of the Secretary of Defense
PA&E	Program Analysis and Evaluation
PAA	primary aircraft authorization
PACAF	Pacific Air Forces
PAF	Project AIR FORCE
PBD 720	Program Budget Decision 720
PDM	programmed depot maintenance
REMIS	Resources Management Information System
RMF	regional maintenance facility
SPG	Strategic Planning Guidance
SSSP	Steady State Security Posture
TAI	total aircraft inventory
TFI	Total Force Integration
UMD	unit manning document
USAFE	U.S. Air Forces in Europe
UTC	unit type code

Introduction

Background

Since the 1990s, the Air Force has been continually engaged in deployed operations in Southwest Asia and in other locations. Recent Office of the Secretary of Defense (OSD) planning guidance directed the services to plan for high levels of engagement and deployed operations, although their nature, locations, durations, and intensity may be unknown. Recognizing that this new guidance might impose different demands on the logistics system, senior Air Force logistics leaders asked RAND Project AIR FORCE to undertake a logistics enterprise analysis.

The operational concepts of recent deployments have shifted, such as partial squadron rotations and deploying to unexpected locations for unknown durations. The deployments have supported a full range of operations, including contingency operations over Serbia, Iraq, and Afghanistan; deterrence operations, such as Southern Watch and Northern Watch; peacekeeping operations; and humanitarian support. The Air Force has recognized that the logistics infrastructure must transform to meet current and future requirements. However, the types of logistics enterprise changes and their extent have not been entirely defined.

The objective of our research was to identify and rethink the basic issues and the premises on which the Air Force plans, organizes, and operates its logistics enterprise in light of these changes. We therefore sought to

- understand how the changes in OSD guidance would affect the Air Force logistics enterprise
- address the allocation of logistics resources strategically to help the Air Force meet the new guidance effectively and efficiently
- explore how the reallocation of workload can enhance enterprisewide logistics capabilities.

Research Motivation

The motivation for this research is to: (a) understand the effects of the changes of OSD guidance on the Air Force logistics enterprise; (b) strategically address the allocation of logistics resources to effectively and efficiently meet the new guidance; and (c) explore how the reallocation of workload can enhance enterprisewide logistics capabilities.

Guidance Changes

The Strategic Planning Guidance (SPG) and the Quadrennial Defense Review for 2000 specified creation of capabilities that will

- ensure homeland defense
- deter aggression in **four** major areas of the world and engage in a number of small-scale contingencies, if needed
- be able to engage in **two** major combat operations (MCOs) simultaneously if deterrence fails in the four areas of strategic importance, and be able to win **one** decisively while engaging in the other until the first is won and then win the second MCO.[1]

The 2004 SPG contained defense planning scenarios to be used for programming operational and support requirements. The scenarios addressed MCOs, a baseline security posture (BSP), homeland security (as part of the Global War on Terrorism), and small-scale contingencies. OSD guidance recognized that the U.S. military would likely be engaged in several global operations at any given time. The guidance also recognized that MCOs, if they occurred, would likely be initiated from an already engaged posture. The guidance instructed the services to size their operational and support forces to execute two MCOs while still providing homeland security, indicating that BSP activities may be curtailed, if necessary, to meet MCO and homeland security requirements.

The 2008 SPG shifted the focus of the military toward irregular, catastrophic, and disruptive threats and capabilities, while maintaining the ability to engage in two MCOs. This guidance replaced the BSP with a set of Steady State Security Posture (SSSP) scenarios.

Looking further into the future, the guidance recognized that the capabilities required of the U.S. military and the roles and missions for each service might change even more dramatically from those the SPG had outlined.[2] Among other future missions, the military services may be asked to

- maintain a substantial and sustained level of effort to suppress terrorist and insurgent groups abroad
- support "hands-on" efforts to train, equip, advise, and assist the forces of nations that seek to suppress insurgents in their own territories
- provide support to defeat internal threats and shore up regional security to cope with external enemies
- overcome modern antiaccess weapons and such methods as theater ballistic missiles and cruise missiles.[3]

Further, redefining roles and missions for the military services and rethinking planning requirements may better prepare each service to respond in the future.[4] For example, given

---

[1] U.S. Department of Defense (DoD), *Quadrennial Defense Review Report*, Washington, D.C., September 30, 2001. The boldface numbers indicate the derivation of the numerical short title for this strategy as the 1-4-2-1 strategy.

[2] See, e.g., U.S. House of Representatives Committee on Armed Services, Panel on Roles and Missions, *Initial Perspectives*, Washington, D.C., January 2008.

[3] Andrew R. Hoehn, Adam Grissom, David A. Ochmanek, David A. Shlapak, and Alan J. Vick, *A New Division of Labor: Meeting America's Security Challenges Beyond Iraq*, Santa Monica, Calif.: RAND Corporation, MG-499-AF, 2007.

[4] Hoehn et al., 2007.

limited resources, ground forces might focus on stability operations, while the Air Force and Navy might focus on large-scale power-projection operations.

In this analysis, we identified the steady-state requirements expected of the Air Force's various mission design series (MDSs). These steady-state requirements generally call for heavy use of intelligence, surveillance, and reconnaissance; mobility; and special operations resources. Fighter and bomber deployment and employment requirements are not expected to be tasked at such high levels outside MCO environments. It may be possible to reallocate resources among some MDSs while ensuring that MCO requirements are met for all aircraft.

Resource Reductions

OSD guidance, such as the SPG, provides the services direction with a perspective on future military operations. The guidance is to be used for force requirements planning and training. Fiscal constraints, however, could inhibit the services' response to the guidance, given existing infrastructure and operating practices. For example, between fiscal years (FYs) 2005 and 2008, Congress cut more than 30,000 positions from the Air Force's end-strength ceiling.[5] In its FY 2009 budget request, the Air Force was planning for a further reduction of 13,000, although Secretary of Defense Robert Gates announced in June 2008 that additional manpower cuts below the FY 2008 end strength were to be put on hold.[6] Program Budget Decision 720 (PBD 720) directed additional manpower reductions, resulting in a total reduction of approximately 57,000 personnel through FY 2011.[7] All these manpower reductions must be achieved without sacrificing the operational capabilities outlined in DoD and Air Force planning guidance. Attrition and manpower savings achieved through base realignment and closure (BRAC) are expected to provide some reductions.

PBD 720–related manpower reductions have affected the Air National Guard (ANG) less dramatically than they do the active-duty force. However, force-structure changes related to the Quadrennial Defense Review and BRAC do potentially affect the ANG. A significant number of legacy aircraft will be retired, many of which are in the ANG.[8] Under current force-employment practices, however, force-structure reductions will not affect ANG end-strength manpower authorizations.

PBD 720–mandated manpower reductions have led to the transfer of activities to contract or civilian organizations. However, PBD 720 manpower reductions have included contractor support personnel. Therefore, contractor support could not be increased as a means of maintaining current capabilities and effectiveness within the Air Force. Taken together, these reductions risk decreases in Air Force capabilities unless it finds ways to maintain effectiveness with less manpower.

---

[5] The Ronald W. Reagan National Defense Authorization Act for Fiscal Year 2005 authorized an Air Force end strength of 359,700; the National Defense Authorization Act for Fiscal Year 2008 authorized an Air Force end strength of 329,563.

[6] In its budget estimate for FY 2009, the Air Force planned for an end strength of 316,600. (Department of the Air Force, "Fiscal Year (FY) 2009 Budget Estimate: Military Personnel Appropriation," briefing, February 2008.)

[7] DoD, Program Budget Decision 720 ("Air Force Transformation Flight Plan"), Washington, D.C., December 2005.

[8] DoD, *BRAC Commission Action Brief*, Washington, D.C., September 1, 2005; DoD, 2001. For example, the BRAC Commission called for the elimination of the flying missions of a number of ANG flying units operating the A-10, F-16, C-130, and KC-135 aircraft.

Research Purpose, Objectives, and Approach

The Air Force recognized that the logistics infrastructure must transform to satisfy this reduction in resources and the Air Force's current and future military requirements simultaneously. The Air Force has initiated a set of interrelated transformation activities, including Air Force Smart Operations 21 and Expeditionary Logistics for the 21st Century, in an attempt to ensure that logistics capabilities evolve in a manner and at a rate compatible with the ongoing evolution of the operational environment.[9] These large-scale activities comprise a broad set of initiatives, including Repair Enterprise–21, centralized asset management, expeditionary combat support system, Air Force Global Logistics Support Center, Air Force Maintenance for the 21st Century, depot "lean" actions, and purchasing and supply chain management. These initiatives are intended to improve the effectiveness and efficiency of Air Force logistics activities.[10]

Senior Air Force leaders asked RAND Project AIR FORCE to undertake a comprehensive strategic reassessment of the entire Air Force logistics enterprise—to reidentify and rethink the basic issues and premises on which the Air Force plans, organizes, and operates its logistics enterprise, from a total force perspective. This analysis addresses the design and strategy of the logistics enterprise by assessing the answers to three questions.

What Will the Logistics Workload Be?

The enterprise strategy will take into account the future force structure, estimated levels of activity, and other factors that influence future workloads and requirements, such as expeditionary deployment and the employment of Air Force assets, from a joint, coalition partner viewpoint.

How Should the Logistics Workload Be Accomplished?

The logistics enterprise can organize and position the workload to support operational requirements effectively and use resources efficiently. Given the enterprise's financial and other resource limits, the strategy must weigh a series of trade-offs, including the following:

"Stockage" and "Response" Solutions. Deployed and engaged forces have historically been supported by a blend of replenishment inventory (readiness spares packages, war readiness engines, prepositioned war-reserve materiel) and support networks (resupply, deployed intermediate-level maintenance, centralized intermediate repair facility support). The decision to use one type of support over another is influenced by in-theater considerations with respect to the size of the force, anticipated usage rates and requirements, the location's accessibility to a resupply network, and the degree to which in-theater units share assets.

Local and Network Maintenance. The maintenance activities that are directly tied to sortie generation in theater or at home station must be supported by aircraft unit-level personnel. However, maintenance activities that are not linked to sortie generation may be supported by locations and organizations other than the aircraft unit. For example, off-equipment component repair can be conducted on base in backshops, or at a number of off-base locations,

---

[9]   Air Force Smart Operations 21 is a method of institutionalizing continuous process improvement in the entire Air Force (including logistics). Expeditionary Logistics for the 21st Century is an effort to transform current logistics processes to provide better support to the warfighter.

[10] Further information on many of these initiatives can be found in U.S. Air Force, Deputy Chief of Staff for Installations and Logistics Directorate of Transformation (AF/A4I), *Logistics Enterprise Architecture (LogEa) Concept Of Operations*, May 24, 2007.

such as Air Force Materiel Command (AFMC) depot facilities, centralized repair facilities (CRFs), or contractor facilities.[11] The geographic component of maintenance activities has strategic implications for the Air Force. For example, a trade-off exists between the reliance on a maintenance and transportation network and the potential manpower savings associated with economies of scale in network-based maintenance operations. The deployment and distribution requirements influence the selection of a local or network maintenance approach for specific types of maintenance.

Contract and Organic Maintenance. The Air Force can partner with contract maintenance organizations to manage non–sortie generation workloads, such as component repair, aircraft inspections, end-item overhauls, aircraft programmed depot maintenance (PDM), and aircraft modification.[12] Moreover, the Air Force can also opt for more-comprehensive contract logistics support arrangements. These decisions can have strategic effects on maintenance and deployment capabilities and may affect Air Force performance and training levels.

Commodity Versus Weapon-System Orientation. The Air Force has historically blended these approaches. Technology repair centers and commodity councils focus on the economies of scope and scale associated with grouping like assets, while system program offices capitalize on the effectiveness trade-offs that focusing on the weapon system as a whole can provide. These organizing constructs have sophisticated strategic nuances, such as efficiency of technology repair centers compared with the simplicity of managing weapon system–unique maintenance networks.

How Should These Questions Be Revisited Over Time?

The Air Force logistics leadership needs to reevaluate the current and future logistics systems' needs and requirements periodically. The allocation of logistics capabilities will shape the strategic design of the Air Force logistics enterprise. To support this decisionmaking process, the Air Force will require experienced logistics analysts who can develop a range of trade-offs for leadership consideration. Moreover, these analysts, ideally within the Air Force, would need to have a clear understanding of the rationale guiding the decisionmaking process to implement the process and periodically execute it.

Answering the Questions

To answer these questions, the Logistics Enterprise Analysis project focused on four major areas. First, we comprehensively evaluated DoD planning guidance to determine projected logistics system workloads. Second, we structurally reviewed scheduled and unscheduled maintenance workloads that may be rebalanced between operating units and support networks. Third, we strategically reevaluated objectives and roles of contract compared with organic support in the logistics enterprise. Fourth, we conducted a top-down review of the management of the existing logistics transformation initiatives to ensure their integrated alignment with broader logistics objectives. A project of this scope is complex, so, to accomplish these objectives, we divided the project into a series of interconnected steps or "spirals."

---

[11] Backshops are one of the three levels of Air Force aircraft maintenance. Historically, *backshops* perform intermediate-level maintenance; *depots* perform the highest level; and *organizational* or *flightline* shops perform the lowest.

[12] It should be noted that for some aircraft, such as the MQ-1 Predator, even sortie-generation workloads have been assigned to contractors.

Focus of This Report

In this report, we concentrate on the portion of our research that addressed aircraft maintenance capabilities between unit-level and network sources of repair for the Air Force fleet of C-130s, including all Air Force–standard C-130Es, C-130Hs, and C-130Js and all specialty variants, including AC-130s, EC-130s, HC-130s, MC-130s, LC-130s, and WC-130s. Other reports address the other dimensions of this research.[13]

We focus on wing-level maintenance tasks, including sortie launch and recovery workloads, aircraft inspections, on-equipment maintenance to support removals and replacement of aircraft components, shop repair of replaceable components, and time-change technical orders. We assume that Air Force personnel will support the workload; however, we discuss an example of contractor support to the C-130 logistics enterprise. In addition, we discuss one transformation initiative that could profoundly affect the sustainment of the C-130 fleet through restructuring the maintenance workload between the wing and depot levels.

Figure 1.1 displays the distribution of the number of aircraft and their locations. The fleet of C-130s is highly distributed with dozens of units supporting limited numbers of aircraft per location. This broad distribution of aircraft is a contributing factor that may make the C-130 a good candidate for rebalancing some of the maintenance workload between the operating locations and a set of CRFs.

A primary goal of this series of analyses is to develop a range of robust alternative policy solutions. Our analytic approach followed four general steps. First, we analyzed defense planning guidance and translated the requirements into a range of potential logistics workloads. Second, we identified what maintenance mission-generation units would conduct at operating locations, both at deployed forward operating locations (FOLs) and at home-station locations. We used this information to determine the manpower requirements needed to support the mission-generation unit. All other maintenance activities were considered candidates for off-site support at a network of CRFs that underpin the mission-generation work centers. Third, we evaluated the performance of the maintenance network focusing on the potential efficiencies associated with the repair capacity, inventory, and distribution system requirements necessary to meet operational objectives. We also examined the effect of centralized maintenance on effectiveness measures, such as aircraft availability. Using an integer linear program, we developed an optimization approach that identifies alternative minimum-cost maintenance network designs, subject to various constraints imposed on the network. Fourth, we used the model iteratively to identify trade-offs of various levels of centralization, from fully centralized solutions to various decentralized solutions. The analysis identifies and compares the alternatives for rebalancing resources invested in the mission-generation work centers with resources invested in the maintenance network. We present a range of options between "enhancing operational effectiveness" and "extracting savings from the logistics enterprise."

---

[13] For example, Ronald G. McGarvey, Manuel Carrillo, Douglas C. Cato, Jr., John G. Drew, Thomas Lang, Kristin F. Lynch, Amy L. Maletic, Hugh G. Massey, James M. Masters, Raymond A. Pyles, Ricardo Sanchez, Jerry M. Sollinger, Brent Thomas, Robert S. Tripp, and Ben D. Van Roo, *Analysis of the Air Force Logistics Enterprise: Evaluation of Global Repair Network Options for Supporting the F-16 and KC-135*, Santa Monica, Calif.: RAND Corporation, MG-872-AF, 2009, and Robert S. Tripp, Ronald G. McGarvey, Ben D. Van Roo, James M. Masters, and Jerry M. Sollinger, *A Repair Network Concept for Air Force Maintenance: Conclusions from Analysis of C-130, F-16, and KC-135 Fleets*, Santa Monica, Calif.: RAND Corporation, MG-919-AF, 2010.

Figure 1.1
Distribution of Standard and Specialty C-130 Aircraft

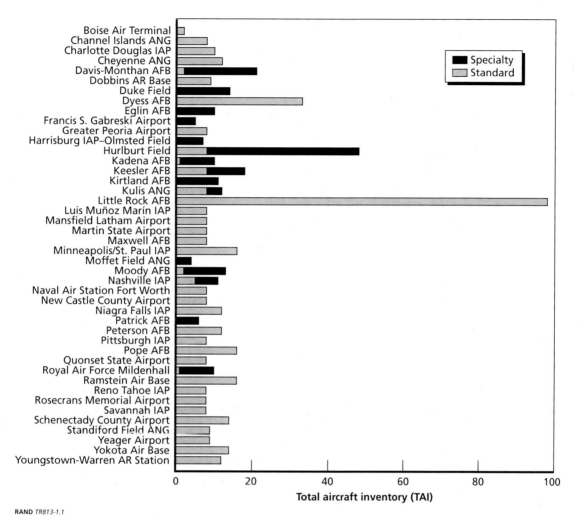

RAND *TR813-1.1*

Our analysis addresses the costs and benefits of an enterprise approach configured to support only active-duty and AFRC forces. We also evaluate an enterprise option that would support the total force.

Organization of This Report

This report has five chapters. Chapter Two details the assessment of logistics workloads associated with OSD planning and programming guidance. Chapter Three presents the results of our maintenance network analysis for the C-130 fleet, while Chapter Four contains our analysis of the integrated maintenance concept, high-velocity maintenance (HVM). Chapter Five presents a summary of our research conclusions. Four appendixes follow Chapter Five and describe our process for determining several key components and calculations. The first addresses manpower authorizations, the second shows how we used the Resources Management Information System (REMIS), and the last two address the models we used.

CHAPTER TWO

Projecting Logistics Workloads

Future deployment, training, and sustainment requirements all substantially affect the logistics system. A workload that is generally predictable is generated by home-station demands, periodic inspections, and overhauls. Deployment operations generate less-predictable workloads. Recent OSD guidance suggests that the demand for deploying airpower will likely remain at the current high level of operating tempo (OPTEMPO) for the foreseeable future. The Air Force must be prepared to support a wide range of operational demands in locations that are difficult to anticipate. Although the Air Force must still ready itself for major conflicts, the nature of such engagements is likely to differ dramatically from the scenarios envisioned in the past. In addition to major conflicts, the Air Force must also remain ready to support national interests in such other operations as peacekeeping, humanitarian assistance, and support of special operations, as recent history has illustrated.

The analysis detailed in this report focuses on wing-level aircraft maintenance for the C-130 fleet. The key data used to identify future workloads are the total number of aircraft operating from each location, the estimated number of combat-coded (CC) and combat-direct support (CA) aircraft in deployed operations, and the total number of flying hours at each aircraft operating location (home station and deployed). Aircraft flying hours are commonly used as a predictor variable to estimate maintenance workload drivers, such as engine failures, while TAI drives scheduled maintenance actions, such as isochronal (ISO) inspections and home-station checks (HSCs).

We identified the home-station beddown for the USAF C-130 TAI, as of the end of FY 2008.[1] We reviewed Program Analysis and Evaluation (PA&E) guidance documents from OSD to identify potential USAF deployment and employment requirements. An important directive within these documents states that services should plan to commence MCOs from a posture in which a significant fraction of the force is already deployed in support of "lesser contingencies." These planning documents present alternatives for force requirements in both the "steady state" of continuous small-scale deployments (i.e., SSSP) and to support MCOs.[2] We based flying-hour requirements on planning factors presented in the War Mobility Plan, Volume 5, using our judgment to adjust these flying hours for different types of deployments (e.g., humanitarian relief operations).[3]

---

[1] Data provided by Maj Jon Julian, AMC/A8PF, in a September 2008 email.

[2] The SSSP has replaced the BSP; DoD and the Joint Chiefs of Staff, *Mobility Capabilities Study*, Washington, D.C., December 19, 2005 (not available to the general public), is in part based on the Baseline Security Posture.

[3] DoD and the Joint Chiefs of Staff, 2005.

Figure 2.1 depicts a notional, but representative, example of how we translated the operational requirements specified in these documents into logistics workloads. The horizontal axis presents time, in years across the Future Years Defense Program (FYDP). Plotted on the vertical axis are the flying hours, for a notional aircraft MDS, projected to be generated over that interval. The sum of the "training" and "SSSP: training offset" areas equals the training flying-hour requirement. Note, however, that only the "training" portion is projected to be flown at the aircraft's home station; in a continuously deployed environment, some of the aircraft expected to be training at home station would deploy to support SSSP operations. The "SSSP: training offset" accounts for these deployed aircraft, capturing their flying hours in SSSP-type operations.[4] The sum of the "SSSP: training offset" and "SSSP: additive requirement" areas equals the total SSSP flying-hour requirement. Note that the extent to which the "SSSP: additive requirement" region is larger than the "SSSP: training offset" region (at any time) indicates the relative difference in OPTEMPO between training and SSSP flying. While the services must support a steady-state deployment requirement, they are also expected to be capable of conducting traditional MCOs. The "MCO" curve presents the dramatic increase in flying-hour requirements associated with conducting such operations.

The OSD guidance is also clear that the services should not plan for any single future but should instead be capable of supporting multiple "strategic environments"[5]; Figure 2.1 attempts to present this concept by showing multiple flying-hour requirement charts for this

Figure 2.1
Notional Flying Hour Requirements

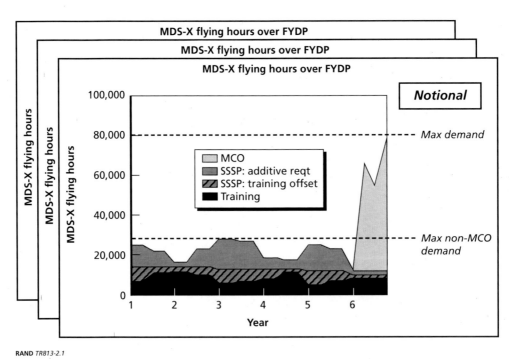

RAND *TR813-2.1*

---

[4]   Note the assumption that the missions flown during a SSSP deployment are perfectly substitutable with training flying requirements. While this may not be the case for all MDSs, we assumed that these hours are a pure offset to training based on discussions with John Cilento, ACC/A3TB, September 2007.

[5]   OSD "strategic environments" may also be referred to as "security environments" in some settings.

MDS. This report presents a method for evaluating ranges of possible futures consistent with these guidance documents, identifying what these alternative futures imply in terms of maintenance manpower requirements.

For our C-130 analysis, we assumed a notional but representative example for the future deployment requirements for standard and specialty aircraft. We assumed that the Air Force requires a capability to support a continuous deployment of 40 percent of the standard CA aircraft and 60 percent of the specialty CC and CA aircraft. A recent snapshot of the Air Force's Logistics Installations and Mission Support–Enterprise View showed approximately 8-percent deployment of the standard fleet.[6] An Air Force Special Operations Command (AFSOC) estimate of the number of deployed specialty C-130s varied between approximately 15 and 30 percent of the specialty fleet.[7]

---

[6] Logistics Installations and Mission Support–Enterprise View snapshot, June 2009.

[7] AFSOC/A4, discussions with authors, June 2009.

CHAPTER THREE

Alternatives for Rebalancing C-130 Maintenance Resources

This chapter describes the process for calculating maintenance-related manpower requirements for the C-130 fleet. In our F-16 and KC-135 analyses, we examined rebalancing alternatives that would either (a) create a new split-operations capability for the fleet; (b) free up manpower resources for other, more stressed, career fields outside aircraft maintenance and capture the savings associated with a reduction in backshop manpower; or (c) combine elements of options (a) and (b). This report presents a similar analysis for the C-130 fleet but goes beyond what we developed for the F-16 and the KC-135 by exploring the potential for reducing the average number of aircraft unavailable due to scheduled maintenance. The reduction of unavailable aircraft is an improvement in maintenance effectiveness.

As with the F-16 and KC-135 analyses, we offer an alternative that rebalances only active-duty and AFRC resources, then extend this analysis to the total force. The chapter begins by describing existing manpower requirements and authorizations. It continues with a discussion of the process of calculating maintenance workloads for peacetime and contingency operations. Next comes an assessment of the inventory pipelines required for supporting centralized operations. The chapter ends by presenting network options for the active-duty and AFRC and summarizes our approach for the total force.

Determination of Maintenance Workload and Manpower Requirements

The first step was to determine the wing-level maintenance authorizations for the current C-130 fleet. We used the FY 2008 year-end unit manning documents (UMDs) for maintenance and maintenance-related positions for the active-duty Air Force, AFRC, and ANG.[1]

Within the following classifications, the active-duty Air Force and AFRC are authorized a total of 13,338 positions: aircraft maintenance squadron (AMXS), equipment maintenance squadron (EMS), component maintenance squadron (CMS), maintenance group (MXG), and maintenance operations squadron (MOS). Our approach for determining manpower totals is discussed in detail in Appendix A, which also contains two tables that provide additional detail

---

[1] RAND's source for UMD manpower authorization data was the end-of-month Manpower Programming and Execution System (MPES) data extract, which we obtained from the access-controlled Air Force MPES website maintained by AF/A1MZ. The MPES data consolidate UMDs for all Air Force organizations and locations in a single table that contains all Air Force manpower requirements, across the total force, including both unfunded manpower requirements and funded manpower authorizations. Our analysis included only funded authorizations and excluded unfunded manpower requirements. Funded authorizations represent the positions in a unit—not the personnel actually assigned. The authorizations are the basis for planning and programming and thus are the most appropriate measure of manpower resources for analytical purposes.

on maintenance manpower for standard and specialty C-130s. Guard authorizations will be discussed later.

Our manpower rebalancing analysis did not address supervisory and support positions from the MXG and MOS classifications across both category types (standard and specialty). However, as with the KC-135 analysis, we did include all maintenance manpower positions in the AMXS, CMS, and EMS except for munitions maintenance, jet engine intermediate maintenance (JEIM), and survival equipment–related maintenance. This analysis assumed that munitions and survival equipment authorizations remain constant and so does not further address or count them. It similarly assumes that all backshop JEIM manpower authorization remain constant. However, the consolidation of JEIM workloads is already under way in the C-130 community.

After estimating the number of authorizations under consideration in the analysis we examined the workload itself, using the Logistics Composite Model (LCOM), a Monte Carlo simulation, to determine the aircraft maintenance resources necessary to support sortie generation objectives. The F-16 analyses used LCOM simulation runs to identify and separate the maintenance workloads that must remain at aircraft operating locations from workloads that could be handled at a centralized off-site facility. The KC-135 analyses instead used Air Mobility Command's (AMC's) existing LCOM analyses and applied the FOL and regional maintenance facility (RMF) concepts to home-station KC-135 operations.[2] AMC has similarly developed C-130 LCOM models to identify the maintenance manpower requirements for home-station operations, deployed FOLs, and RMFs supporting aircraft at deployed FOLs.[3]

Table 3.1 categorizes tasks from AMC's C-130 RMF/FOL construct into the following general workload categories at the FOLs and at the home station: mission generation, scheduled maintenance, deferrable maintenance, and off-equipment maintenance. The home stations conduct their own maintenance in all four categories. Mission-generation tasks support operations at both FOLs and home stations. These workloads must therefore remain at the aircraft operating locations. In the AMC RMF/FOL concept, the FOLs perform their HSCs, and centralize their ISO inspections to RMFs. The AMC concept does not consider the refurbishment process for FOLs. Deferrable maintenance (nongrounding failures [NGFs]) and off-equipment component repair could take place at any location, although there is an assumption that risk tolerance is greater at locations that do not perform their own NGF work. AMC's RMF/FOL concept sends both workloads off site, from the FOL to the RMF, accepting this NGF risk at forward-deployed locations in exchange for reducing the on-site maintenance resources deployed to the FOL.

We examined extending AMC's RMF/FOL concept in an attempt to rebalance workloads and resources between FOLs, home-station units, and a network of CRFs. We first identified the workloads that could be centralized, then used LCOM to isolate the workloads and identify any manpower economies of scale that centralization might realize.

We began by examining major scheduled maintenance tasks, including ISO inspections, HSCs, and refurbishments. C-130s undergo an ISO inspection every 450 days; HSCs take

---

[2]   Shenita L. Clay, *KC-135 Logistics Composite Model (LCOM) Final Report: Peacetime and Wartime—Peacetime Update*, Scott AFB, Ill.: HQ AMC/XPMMS, May 1, 1999.

[3]   Headquarters Air Mobility Command, *C-130 Logistics Composite Model (LCOM) Final Report: Peacetime and Wartime*, Scott AFB, Ill.: HQ AMC/XPMRL, February 12, 2001. Note that this analysis did not utilize LCOM models for non-AMC versions of the C-130.

Table 3.1
AMC RMF/FOL C-130 LCOM Workload Allocation

Location of Aircraft	Mission Generation			Scheduled MX		Deferrable Nongrounding Failures	Off-Equipment Component Repair
	Launch and Recover	Remove and Replace	Home-Station Checks	Isochronal Inspection	Refurbishment		
FOL					N/A		
Home Station							

Workload at aircraft location
Workload at RMF

place during the calendar midpoint between ISO inspection cycles; and the refurbishment process (currently considered optional) occurs every 36 months. These inspections vary in purpose and maintenance resource requirements. The HSC is the least time- and labor-intensive process and primarily consists of a general, noninvasive structural inspection of the aircraft. Crew chiefs perform the majority of the required tasks and can do so either on the flightline or in a hangar.

The ISO inspection is more invasive but still centers on the structural aspects of the aircraft. It is often performed in a hangar to accommodate removing and replacing aircraft panels. Although it is a structural inspection by nature, a representative workload may be required by non–crew chief Air Force Specialty Codes and backshops.

Finally, the refurbishment process is often the most labor-intensive process, although the Air Force considers it optional because many tasks involved are not critical to aircraft readiness or availability. These noncritical tasks include interior and exterior improvements, such as webbing replacement and floorboard repair.

Scheduled Maintenance

Isochronal Inspections. Using REMIS data for a five-year period (January 2002 to January 2007), we calculated the ISO inspection "fly-to-fly" times, the maintenance man-hours (MMH) associated with an ISO inspection, and the ISO inspection throughput times.[4] On average, it took approximately 2,500 MMH to complete an ISO inspection. The average fly-to-fly time was 40 days, while the average throughput time was approximately 15 days.[5]

Figure 3.1 further illustrates one of the calculations, fly-to-fly time, by major command (MAJCOM). The x-axis presents the number of fly-to-fly days, and the y-axis reflects the probability distribution function of the fly-to-fly times falling into each specific period. Three important points emerge from Figure 3.1 and our analysis of ISO inspection maintenance data. First, the ISO inspection process is time and labor intensive, and there can be considerable uncertainty in the fly-to-fly interval and the MMH needed to complete the ISO inspection process.[6] The ISO inspection fly-to-fly interval is particularly important because it can be translated into an estimate of the number of aircraft that are unavailable because they are in the ISO inspection process. Aircraft availability is an important consideration for the C-130 because this aircraft is so heavily tasked in both the current and projected future operating environments.

Second, average fly-to-fly times differ across MAJCOMs. The variations do not necessarily indicate performance discrepancies. Rather, they highlight the staffing of ISO inspection–related work centers according to the number of aircraft the location supports. Units that support a larger number of aircraft have more people and often more shifts in the work centers to handle the additional ISO inspections than do units with a smaller number of aircraft. Therefore, the fly-to-fly period decreases per ISO inspection as unit size increases.

---

[4]   Appendix B describes these calculations.

[5]   This average throughput time accounts for and eliminates the downtime occurring while the aircraft is in work (e.g., a weekend when no maintenance takes place).

[6]   See Appendix B for a detailed listing of the mean and standard deviations of the fly-to-fly, MMH, and flow times of ISO inspections. A minor component of the variability in the ISO inspection process can be attributed to the two different ISO inspection work packages used over a span of four cycles (three minor ISO inspections and a major ISO inspection).

Figure 3.1
C-130 Distribution of Fly-to-Fly Times Organized by MAJCOM

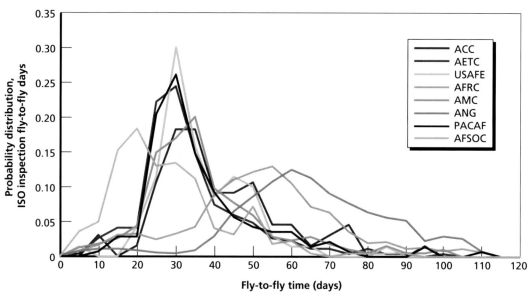

RAND *TR813-3.1*

Third, Figure 3.1 provides a general sense of how the variability associated with the ISO inspection process may make it inefficient to perform ISO inspections at the unit level. Although the fly-to-fly days vary among MAJCOMs, the actual flow days, or calendar time spent working on the aircraft, are comparable across MAJCOMs. If work-center staffing policies were similar across the MAJCOMs, the ISO inspection fly-to-fly times would also likely be similar.

Although staffing policies can reconcile some of the variability of ISO inspections, the variability that remains raises the question of where the work should take place. An operating location with fewer aircraft may have scheduled maintenance workloads with a high degree of variability in the inspection and repair process. This variability may make the workload difficult to plan for and difficult for maintenance personnel to absorb. Smaller operating locations can either employ large amounts of underutilized maintenance resources to manage the variability or may encounter some long queuing maintenance repair times when the workload nears or exceeds repair capacities. Large units or CRFs benefit from the phenomenon that, as workloads are aggregated across locations, the magnitude of the variability relative to the planned repair capacity will likely not be as pronounced. In addition, the maintenance personnel at the large facilities are better positioned to manage the aggregated workload without employing large reserves of underutilized resources.

Because ISO inspections are labor intensive and because the required flow times and MMH vary significantly, they are a strong candidate for centralization. We modified the LCOM simulations to build one that evaluated the performance of a CRF that performs only ISO inspections. Figure 3.2 illustrates the economies of scale associated with centralization of the ISO inspection process. The horizontal axis indicates the capacity for annual ISO inspections at a site, and the vertical axis shows the maintenance manpower required per ISO inspection per year. The figure shows that a relatively small facility supporting 13 ISO inspections per year requires approximately five maintenance manpower authorizations per ISO inspection

Figure 3.2
C-130 Economies of Scale in the ISO Inspection Process

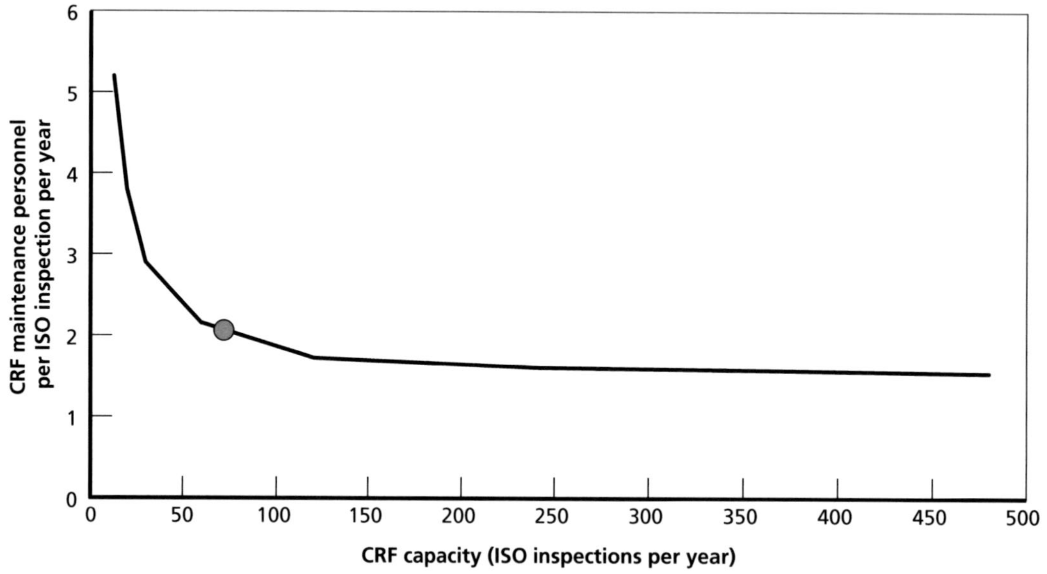

NOTE: The dot on the curve indicates the actual manpower level for Little Rock.

RAND *TR813-3.2*

per year. As the inspection capacity of the CRF increases, the proportional number of manpower authorizations it requires declines sharply, leveling off at just below two authorizations per ISO inspection per year, at the 250-per-year facility. Beyond that point, the curve flattens, demonstrating very little additional marginal reduction in manpower for facilities with larger CRF capacity.

The observed economies of scale are significant and arise because larger maintenance operations can achieve greater personnel utilization. However, because the Air Force's experience with LCOM has been limited to the smaller unit sizes at the left end of this curve, reservations about the accuracy of LCOM at the large unit sizes, which involve a significant extrapolation of the model's domain, may be warranted. Fortunately, we were able to compare the LCOM estimates for the C-130 analysis with an existing large-scale ISO inspection operation to gauge the validity of the LCOM model for large units. Little Rock Air Force Base (AFB) has dedicated manpower and facilities to an ISO inspection–only maintenance operation that completes approximately 75 ISO inspections each year. We used Little Rock's FY 2008 actual filled manpower positions for comparison, represented by the circle in Figure 3.2. Note that the actual manpower value falls directly on the LCOM curve.[7] More important, the fact that Little Rock's actual manpower is able to accomplish the required workload demonstrates that the manpower levels LCOM suggests are adequate for the required workload, at least up to a facility size of 75 annual ISO inspections. Moreover, this value lies at a point along the curve that captures very large reductions in the proportional manpower required relative to that

---

[7] Note that Air Education and Training Command (AETC), which determined the FY 2008 authorizations, does not use a pure LCOM-based approach. Thus, the agreement between the authorized level and our LCOM results is not simply a self-fulfilling prophecy.

for small operations. These data suggest that the estimated economies of scale present in the LCOM model accurately reflect existing large-scale facilities.

Figure 3.3 illustrates the maintenance manpower utilization for the ISO inspection maintenance process. Smaller, decentralized maintenance operations utilize less of their available manpower because of (1) minimum-crew-size effects, which are driven by the work-center task that requires the largest crew, even if most work-center tasks require smaller crews, and (2) "insurance" effects, which mandate that the organization have the capacity to accommodate random spikes in demand without a large negative effect on flying operations.

Conversely, centralized maintenance operations with high-volume workloads better utilize their manpower for two reasons. First, pooling workloads reduces minimum-crew-size effects. Because the total manpower requirement for a shop with a large maintenance requirement will be larger than the minimum crew size, its manpower can be sized in accordance with the workload, allowing for higher manpower utilization. Second, higher work volumes decrease the manpower insurance requirements often associated with small but variable workloads. Both the workload demand and the duration of maintenance activities fluctuate randomly, so a manpower utilization close to 100 percent would mean significant queues of unavailable components and aircraft. In fact, smaller maintenance operations must maintain quite low manpower utilizations (20 percent, on average, for the 13 ISO inspection per year CRF; see the left endpoint of Figure 3.3), independent of minimum-crew-size effects, to ensure that adequate capacity is available to accommodate spikes in workload requirements and repair durations.[8] Larger maintenance operations utilize more of their manpower (58 percent, on average, for the 480 ISO inspection per year CRF; see the right endpoint of Figure 3.3) without generating

Figure 3.3
Manpower Utilization for the ISO Inspection Process

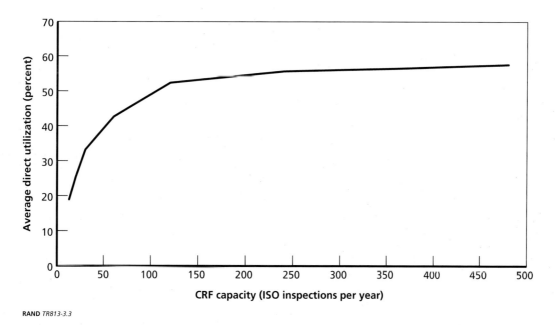

---

[8] We assumed that each LCOM work center could not exceed an average of 60 percent utilization. We then further adjusted our calculations of LCOM work-center requirements accordingly, incorporating man-hour availability factors (MAFs).

large queues of unavailable components and aircraft. These operations can do so because of the decreasing coefficient of variation of the workload and repair times that pooling demand and repair resources provides.

In our earlier F-16 and KC-135 analysis,[9] we assumed that CRFs would be staffed at the level necessary to support a fixed flow time through the process. This assumption forced increasing manpower at smaller CRF facilities because they had to support round-the-clock, seven-day-a-week operations to maintain the desired flow time. For the C-130 analysis, we relaxed the CRF flow-time assumption. Instead, consistent with current Air Force practice, second and third maintenance shifts would come in only to support the workload requirement. This meant that a smaller CRF might staff only some work centers with one to two shifts and, as the size of its operations grew, would increase the number of personnel and shifts. As a result, the ISO inspection flow times are now a function of the size of the CRF operations. Flow times would be longer for smaller CRFs than for larger ones able to support round-the-clock, seven-day-a-week operations. Figure 3.4 shows the relationship between flow times and CRF capacity that derive from our LCOM analyses. For example, a CRF supporting 13 ISO inspections per year could expect an average ISO inspection flow time of approximately 25 days; a CRF supporting 60 ISO inspections per year could expect a flow time of 12 days.

We again used empirical data points to validate the estimated LCOM model flow times for the ISO inspection process against actual values for two large-scale ISO inspection facilities: Little Rock AFB and Hurlburt Field, which acts as a CRF for AFSOC aircraft. These operations are indicated in Figure 3.4 as a circle and square, respectively. Given their relative size, the flow times of these operations compare favorably with the LCOM estimates and again appear to validate the LCOM model outputs. Little Rock staffs two to two-and-a-half shifts per day and does not always work multiple-shift operations on the weekends. This explains the average flow days being slightly larger than LCOM estimates for its facility size. AFSOC, which employs a contractor to perform the ISO inspection process, primarily uses three-shift operations.

Figure 3.4 also shows how CRFs can help improve effectiveness for the Air Force. The C-130 network consists of 22 active-duty and AFRC locations. More than one-half of these perform fewer than ten annual ISO inspections, and only three perform more than 20. Because a majority of locations perform only a limited number of ISO inspections per year, ISO inspection–related work centers are not necessarily staffed to support multiple shifts, and the ISO inspection process, when coupled with other tasks, can exceed 40 days per ISO inspection per aircraft.[10] In a typical squadron, this may mean that one or two aircraft are always unavailable to the operator. Across the active-duty and AFRC fleet, this could translate into more than 30 aircraft (approximately 10 percent) on average being in the ISO inspection process at any one time. If the active-duty and AFRC network implemented the CRF process, the average number of aircraft in the ISO inspection process would decrease to less than 13. The gain in aircraft for the fleet may not be needed if the fleet requirements during steady state and surge operations are low and enough aircraft already exist. However, the effectiveness gains are critical for such fleets as the standard and specialty C-130s because they are in high demand. Other

---

[9] McGarvey, Carrillo, et al., 2009.

[10] As noted earlier, the average ISO inspection fly-to-fly time is approximately 40 days. This estimate captures ISO inspection and ISO inspection–related and other maintenance performed during the ISO inspection fly-to-fly period.

Figure 3.4
ISO Inspection Flow Times as a Function of Size of CRF Operations

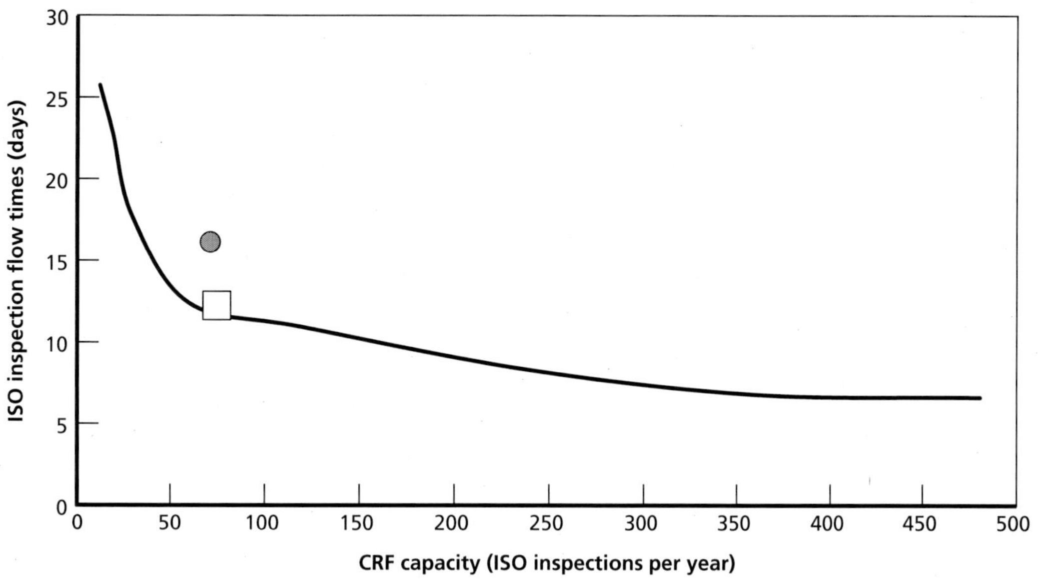

NOTE: The square indicates AFSOC; the dot, Little Rock.

RAND TR813-3.4

high-use fleets may consider the CRF concept as a means of decreasing the number of aircraft in scheduled maintenance processes.

Consolidating the workload to CRFs would likely also lead to reductions in flow time variability associated with supply shortages. Similar to pooling workloads, large CRF operations would benefit from pooling the demand and the supply of components and materials used in the ISO inspection process. Conversely, as CRFs transition from small one-shift operations to two- and three-shift operations, the importance of available supply increases. Larger CRFs may not fully realize manpower efficiencies if lengthy supply delays are incurred.

Our analysis leverages AMC's LCOM model, which assumes that maintenance manpower resources are not limited by supply availability issues. The model accounts for line-replaceable unit failures in the repair process, which properly captures repair flow times. It does not, however, capture inventory levels and non–mission capable due to supply. The severity of parts availability issues will likely be somewhat limited because the ISO inspection process mainly comprises inspections and scheduled maintenance.

Home-Station Checks. We considered the HSC process to be another scheduled maintenance task that, if centralized, could improve efficiency and effectiveness. We repeated an empirical examination of HSCs using REMIS (January 2002 to January 2007).[11] HSCs took considerably less time and manpower than ISO inspections, but the data shared some of the general characteristics. It took an average of approximately 375 MMH required to complete an HSC. The average fly-to-fly time was 14 days, while the average throughput time was approximately four days.[12] As previously mentioned, HSCs occur in the middle of the ISO inspection time cycle—225 days after the last ISO inspection. We again modified the LCOM

---

[11] Appendix B discusses our approach.

[12] Removing the downtime—the time when no maintenance was conducted.

simulations, this time isolating the manpower requirement associated with HSC. A centralized HSC facility could capture some economies of scale by supporting a large number of HSCs. The manpower resource requirement necessary to support a centralized HSC facility was lower than that required for a centralized ISO inspection facility. LCOM runs without the HSC workload from the FOLs and the home stations did not show a substantially lower total workload at the operating locations. HSCs account for only 1 to 3 percent of the FOL workload, which is sufficiently small that the manpower authorizations at the operating locations did not decrease when the HSC workload was removed. Therefore, the centralization of the HSC process would add to the total system manpower bill because of the CRF manpower positions needed to support large-scale HSC operations. Shuttling and facilities costs would also be included. These things mean that centralized HSCs would not be cost-effective, and we thus excluded them from the candidate workloads for centralization.

However, there may be other reasons to centralize some HSCs or not to do them at specific operating locations. In a deployed operating environment, weather, or a lack of facility space may make conducting HSCs at FOLs undesirable. In such situations, the centralization of HSCs at a site (potentially, in theater) with an existing facility might be desirable. Note, however, that other options are available for avoiding the requirement to perform HSCs at austere FOLs. If rotational policies for deployed aircraft were shorter than the HSC intervals, units could rotate aircraft back to their home stations to perform HSCs.[13] Such a policy would require significant planning and scheduling capabilities at the units (or at a centralized organization, if the replacement aircraft were not necessarily sourced from the same unit).

Refurbishment. We examined other workloads that could be conducted at the CRF. We examined the 36-month refurbishment process as another scheduled maintenance task that could be centralized. The refurbishment process is officially optional. However, discussions with personnel at the operating locations suggested that, even though refurbishments are optional, the refurbishment-based workload was generally conducted in line with the ISO inspection process. Therefore, to account for this additional workload, we adjusted our LCOM model so the current 36-month refurbishment interval would align with every second ISO inspection (moving to a 900-day interval). In addition, we conservatively assumed the refurbishment workload and flow time would add to the CRF ISO inspection process.[14] Figure 3.5 compares the economies of scale associated with AMC's RMF concept, in which the centralized facility performs ISO inspections but not refurbishments (as presented in Figures 3.2 and 3.4), and RAND's CRF concept, in which CRFs perform both. Figure 3.6 compares the flow times of AMC's RMF concept with RAND's CRF concept.

Deferrable Maintenance

In addition to scheduled maintenance tasks, we examined centralizing deferrable maintenance associated with NGF write-ups. This workload is a function of the sortie-generation and FOL flying requirements. Centralizing this workload did not generate economies of scale and

---

[13] We examined a subset of data beginning in July 2007 through February 2008 from 19 AMC deployed C-130s on the average deployment length of an aircraft. The aircraft were deployed for an average of approximately 200 days, with a standard deviation of 64 days. Some aircraft underwent HSCs while deployed.

[14] Some refurbishment workloads could be synchronized with the existing ISO inspection workload to reduce the estimated flow times.

Figure 3.5
Economies of Scale: AMC's RMF and RAND's CRF Concepts

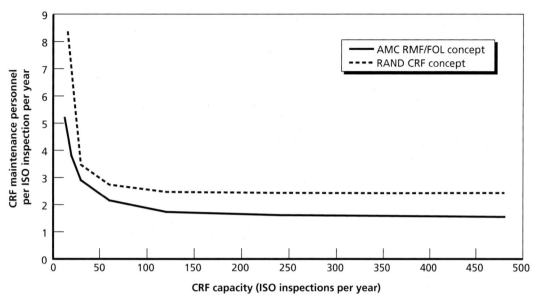

RAND *TR813-3.5*

Figure 3.6
Flow Times of AMC's RMF Concept and RAND's CRF Concept

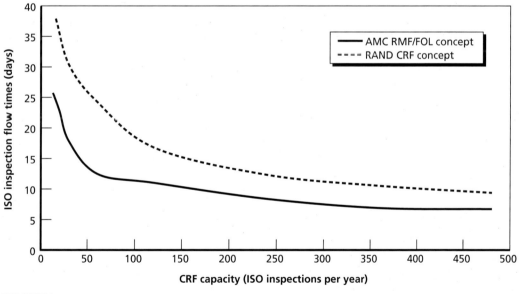

RAND *TR813-3.6*

increased in a near-linear function with the number of aircraft supported at an FOL.[15] This means that the NGF workload could be conducted at either the CRF or at the home station. As noted above, AMC's RMF/FOL concept sends this NGF workload off site from the FOL

---

[15] The manpower requirement that supports NGF workloads from FOL operations is determined by the deployment assumptions.

and on to the RMF, accepting the increased risk at forward-deployed locations in exchange for reducing the on-site maintenance resources deployed to the FOL. We assumed that home stations would not accept this risk and subsequently assumed that they would perform the NGF workload generated at the home station, while the FOL-related NGF workload would be conducted by a mixture of manpower at the CRF and at home station.

Off-Equipment Component Repair

We next determined whether centralizing the off-equipment component repair for both the home station and the FOL could offer additional economies of scale. AMC's FOL/RMF LCOM models assume that the CRF supports all FOL off-equipment workload.

Off-equipment shops include guidance and control, communication and navigation, pneudraulics, electric, and environmental (at both the FOL and home station); and metals technology (at the FOL), in addition to a limited set of JEIM workload (at home station).[16] We observed additional economies of scale from centralizing some of the off-equipment workload. Table 3.2 summarizes the allocation of workloads between operating locations and the CRF. CRFs support all operating locations' ISO inspections, refurbishments, all FOL off-equipment component repair. The home-station locations support the home-station NGF workload, while a combination of home station and CRF facilities would support the FOL NGF workload.

Mission-Generation Maintenance

Given this allocation of maintenance workloads, we next needed to determine the maintenance manpower required to perform the mission-generation tasks at FOLs and home station. The process for determining manpower requirements differed slightly for standard and specialty aircraft.

Because AMC has incorporated FOL maintenance into its deployment concepts for standard C-130s and because our analysis did not change the set of tasks to be conducted at the FOLs, it was not necessary to perform new LCOM runs to identify these requirements. Instead, we used the unit type code (UTC) deployment manpower requirements for C-130 maintenance and applied them to all CA squadrons.[17]

It is difficult to apply the UTC approach to determine home-station manpower requirements for non-CA squadrons. The home-station maintenance construct described above differs slightly from the FOL construct. Since this new home-station maintenance construct also departs from current Air Force practice for standard C-130s, new LCOM analyses were neces-

---

[16] We did not examine centralization of all JEIM workloads. Rather, only the limited set of JEIM tasks included in the AMC home-station LCOM model was assumed to be sent to the CRF. However, our discussions with operating locations indicate that an aircraft undergoing the ISO inspection process creates a significant engine workload. A common location for CRFs and JEIM centralized intermediate repair facilities may lead to manpower and component shuttling savings.

[17] The following UTCs were used for this analysis: for the C-130E, HNE4L, HNE41, and HNE22; for the C-130H, HNH4L, HNH41, and HNH22; and for the C-130J, HNJ4L and HNJ41. UTC positions identified as belonging to the MOS, supply squadron, and command post were excluded from this analysis. A number of the UTC maintenance work centers (flightline crew chief, flightline communications and navigation, flightline guidance and control, flightline electronic countermeasures, flightline propulsion, flightline hydraulics, flightline electrics and environmental) are presently organized under the AMXS. The remaining UTC work centers (metals technology, nondestructive inspection, structural repair, fuels, aerospace repair, aircraft ground equipment) are currently organized under the CMS and EMS. Note that these CMS and EMS work centers would also be required at a CRF, requiring these shops to split between the aircraft operating locations and CRFs.

Table 3.2
RAND CRF C-130 LCOM Workload Allocation

Location of Aircraft	Mission Generation		Scheduled MX				
	Launch and Recover	Remove and Replace	Home-Station Checks	Isochronal Inspection	Refurbishment	Deferrable Nongrounding Failures	Off-Equipment Component Repair
FOL							
Home Station							

Workload at aircraft location
Workload at CRF
Workload at mixture of locations

sary to identify the maintenance manpower for the mission-generation workloads at home station. The AMC C-130 home-station LCOM model was modified to remove ISO inspections and aircraft refurbishment. Furthermore, we assumed that every fifth aircraft wash would be conducted at a CRF (since aircraft are washed every 90 days and are assumed to have their 450-day ISO inspections conducted at a CRF).

The AMC LCOM model identifies separate manpower requirements for flightline maintenance specialists (in communications and navigation, electrics, and environmental, guidance and control, hydraulics, and propulsion) and the backshops that perform off-equipment maintenance in these same specialty areas. Our analysis assumed that this set of off-equipment workloads would be reassigned to CRFs, thus the backshop manpower associated with these specialty areas was not included in the requirements for the mission-generation work centers.

The C-130 UTCs do not include the aircraft inspection and wheel-and-tire work centers. However, because we assumed that some aircraft inspection tasks, such as aircraft wash, would still be conducted at home station, and that home-station units would maintain a wheel-and-tire shop, it was necessary to include these work centers in the home-station LCOM model and determine a manpower requirement for these mission-generation work centers, even though these shops have no FOL requirement. The C-130 LCOM model also does not include the nondestructive inspection and aircraft ground equipment (AGE) work centers, so the manpower requirements for these mission-generation work centers came solely from the UTCs. This analysis assumed that the munitions flight would not be modified under these new maintenance constructs and so excluded personnel in the munitions flight.[18]

Table 3.3
Distribution of Mission-Generation Maintenance Work Centers for Standard C-130s

Mission-Generation Work Centers	CRF Work Centers
Crew chief—flightline	Aero repair—CRF
Communications and navigation—flightline	APG inspection—CRF
ECM—flightline	Fuels—CRF
GAC—flightline	Metals tech—CRF
Propulsion—flightline	Structural repair—CRF
Pneudraulics—flightline	NDI—CRF
E&E—flightline	Wheel and tire—CRF
Aero repair—mission generation	Communications and navigation
APG inspection—mission generation	ECM
Fuels—mission generation	GAC
Metals tech—mission generation	Propulsion
Structural repair—mission generation	Pneudraulics
NDI—mission generation	E&E
Wheel and tire—mission generation	
AGE—mission generation	
Munitions—mission generation	

NOTE: The shaded areas indicates shops whose workloads are split between the mission-generation work centers and the CRF.

---

[18] These personnel do not appear in any of our manpower counts.

Table 3.3 presents the set of work centers that would be associated with mission-generation work-center maintenance and CRF maintenance for the standard C-130s. The workloads of the highlighted shops are split between the mission-generation work centers and the CRF.

We observed that the ability to conduct split operations, wherein C-130 squadrons deploy some fraction of their primary aircraft authorization (PAA) but also leave some of it at home station, is consistent both with planning and programming guidance and with recent experience. In general, this concept and deploying partial squadrons creates diseconomies of scale and requires additional manpower—both air crews and maintainers.

The additional manpower requirement was determined from a scenario in which all CA squadrons were to be resourced in accordance with a split-operations capability. For a representative 16 PAA CA squadron, we assumed a split operation: Eight aircraft would deploy with the UTC-specified manpower; the other eight aircraft would remain at home station, with the manpower complement RAND's home-station LCOM model specified. We then computed a value we called the "split-operations plus-up"—the difference between the manpower requirement for a split-operation and the UTC requirement for deploying the entire squadron.

These requirements appear in Table 3.4. The top row shows the FY 2008 manpower authorization totals for standard C-130 AMXS. The second and third rows are the mission-generation unit manpower numbers. These numbers are based on a UTC-based requirement for CA squadrons and an LCOM-based requirement for non-CA squadrons. The rows differentiate between manpower formerly assigned to the AMXS and manpower that is formerly assigned to the CMS and EMS (which would be operating as split shops). The fourth data row presents the requirement for a split-operations plus-up at all CA squadrons. The C-130 split-operations capability requires approximately 3,500 additional maintenance manpower positions beyond those required for each squadron to operate in a fully deployed scenario.

The process for determining the maintenance manpower requirements for mission-generation work centers for specialty C-130s was somewhat different. As with the standard C-130s, we used the UTC deployment manpower requirements for aircraft maintenance and applied them to all CC and CA squadrons.[19] However, the UMD flightline maintenance manpower in AFSOC CC and CA units was generally twice as large as the unit's UTC requirement. The difference is due to the extremely high frequency with which AFSOC aircraft deploy. Thus, when computing a UTC-based deployment requirement for AFSOC aircraft, we assigned to each CC and CA unit a manpower level equal to two times its UTC requirement. Because we did not perform any LCOM runs for specialty C-130 aircraft, for AFSOC non-CC, or for non-CA units, we assumed that the home-station manpower requirement would equal the UTC requirement for a similarly sized unit. Because the CC and CA AFSOC units were to be twice their UTC requirement, we assumed that no further split-operations require-

---

[19] This analysis used the following UTCs: for the AC-130H, HSAWH and HSAWG; for the AC-130U, HSAU3, HSAU4, and HSAU6; for the EC-130J, HSAXH and HSAXJ; for the MC-130E, HSAGB; for the MC-130H, HSSM3, HSSM4, and HSSM6; for the MC-130P, HSSP2, and HSSH4; for the MC-130W, HSSC1; for the EC-130H, HFCAL, HFCA1, and HFCA2; for the HC-130, HFRAL, HFRA1, HFRA2, and HFRA3; for the LC-130H, HNL4L, HNL41, and HNL22; for the MC-130 (CSAR, non-AFSOC), HFRML, HFRM1, and HFRM2; and for the WC-130J, HWJWL, HNJ4L, HWJW1, HNJ41, HNJW2, and HNJ22. Note that UTC positions identified as belonging to the MOS, supply squadron, and command post were excluded from this analysis. These data were obtained from the AF Portal site along the following path: *FOA : AFMA–Air Force Manpower Agency : MAS–Requirements Support Division : Wartime Readiness : MANFOR in August 2008.*

ment was warranted at AFSOC units; these units received a manpower increase beyond their UTC requirement in the non–split-operations baseline.

The specialty C-130s have a number of additional UTC work centers not found in standard C-130 units. For example, the AC-130H and AC-130U have a UTC work center for armaments, while the EC-130J has a UTC work center for ground radio communications. The mission-generation work-center constructs for specialty C-130s included these additional UTC work centers. However, as with the standard C-130, this analysis excluded personnel in the munitions flight, and so they do not appear in our manpower counts.

For non-AFSOC specialty C-130s, the manpower for each unit's mission-generation work centers was sized according to its UTC requirement, much as for the standard C-130s. The split-operations plus-up for such units was equal to an additional UTC's worth of manpower, since we did not perform LCOM runs to determine specialty C-130 home-station manpower requirements. Note that this produces a split-operations requirement that is larger for non-AFSOC specialty aircraft (twice the unit's UTC requirement) than that for standard aircraft (UTC requirement for one-half the PAA plus home-station requirement for one-half the PAA). We resourced the split-operations manpower for non-AFSOC specialty units at a higher level than for the standard units because of the heavier deployment burden identified for these specialty aircraft.

Table 3.4
Manpower Requirements for Standard C-130 Mission-Generation Operations

Standard C-130	Active-Duty	ANG		AFRC		Total
		Part-Time	Full-Time	Part-Time	Full-Time	
AMXS FY08 UMD	2,418	703	462	906	288	4,777
UTC-based						
AMXS	1,702	1,178	854	636	392	4,762
Moved from CMS and EMS	537	341	247	186	114	1,425
Split-operations plus-up	780	1,087	788	533	329	3,517
Proposed new mission-generation work center	3,019	2,606	1,889	1,355	835	9,704

Table 3.5
Manpower Requirements for Specialty C-130 Mission-Generation Operations

Specialty C-130	Active-Duty	ANG		AFARC		Total
		Part-Time	Full-Time	Part-Time	Full-Time	
AMXS FY08 UMD	2,663	152	123	210	87	3,235
UTC-based						
AMXS	2,003	171	157	269	171	2,771
Moved from CMS and EMS	969	208	190	129	82	1,578
Split-operations plus-up	482	188	172	166	105	1,113
Proposed new mission-generation work center	3,454	567	519	564	358	5,462

Table 3.5 presents the manpower requirements that resulted from this analysis for specialty C-130s. This specialty C-130 split-operations capability requires approximately 1,100 additional maintenance manpower positions.

C-130 Component Repair Pipeline

The off-equipment component repair workload removed from operating locations and sent to CRFs would require inventory to support the transportation and repair pipeline and would generate a transportation bill for shipping inventory between the locations and the CRF. This analysis focused on the set of aircraft components appearing in both (a) the current readiness spares package for any USAF C-130 unit and (b) the RAND March 2006 data drawn from the D200 requirements data bank. Across all variants, this intersection comprises a set of 870 unique national item identification numbers (NIINs). This approach is similar to our pipeline analysis for the F-16 and KC-135.[20]

The inventory requirement associated with CRF work centers performing operating locations' component repair workload is limited to component failures that were previously repaired at the on-site backshops. We estimated this effect by multiplying the expected number of component failures by the percentage of base not reparable this station (BNRTS). We assume a notional home-station daily flying schedule of 2.5 flying hours per PAA for the set of NIINs. Using this subset of NIINs, and limiting the analysis to failures currently repaired on site, we would expect to observe a daily fleetwide average of 96.8 component failures. To estimate the transport costs associated with the use of CRFs in support of home-station operations (including operations for permanently assigned Pacific Air Forces [PACAF] and U.S. Air Forces in Europe [USAFE] forces), we assumed that all failed components would be shipped using FedEx Small Package Express two-day rates for U.S. domestic shipments.[21] Focusing solely on workloads formerly conducted in the backshops for the limited number of work centers under consideration, the expected annual fleetwide transportation cost to support home-station operations is approximately $2.9 million a year.

An inventory requirement can be similarly computed. As with the transportation computations, additional inventory would be necessary to support the new transportation segments the CRF would introduce. Using the two-day transport time assumed above, in each direction, generates a requirement for four days' worth of pipeline inventory. Computing a separate inventory requirement to support each of the permanently assigned USAFE, PACAF, and continental United States (CONUS) C-130 fleets, operating at a notional 2.5 flying hours per PAA, results in an additive requirement amounting to a $19.7 million investment. This inven-

---

[20] For additional information on the technique developed to evaluate the component pipeline, please see Appendix F of Ronald G. McGarvey, James M. Masters, Louis Luangkesorn, Stephen Sheehy, John G. Drew, Robert Kerchner, Ben Van Roo, and Charles Robert Roll, Jr., *Supporting Air and Space Expeditionary Forces: Analysis of CONUS Centralized Intermediate Repair Facilities*, Santa Monica, Calif.: RAND Corporation, MG-418-AF, 2008.

[21] Federal Express (FedEx) rates were used as an example of a potential carrier. RAND does not endorse the use of FedEx or any other carrier. We obtained information on applicable FedEx rates from the Government Services Administration website.

tory investment is a one-time cost that could be amortized over the expected duration of the CRF option. The $19.7 million is a requirement against the actual demand.[22]

C-130 Repair Network Design Options

We begin the network analysis by focusing on the requirements of the contingency-deployed fleet. We considered only one option for providing CRF support to the C-130 deployed fleets: Aircraft and components are retrograded to a fixed network site, which might be at a permanent PACAF or USAFE installation. As with the previous KC-135 analysis,[23] this one does not include a forward-deployed CRF capability in the C-130 analysis. The C-130's flying range and long, calendar-based maintenance intervals suggest that establishing a contingency CRF should not be necessary.

The optimization approach is similar to the approaches for our earlier F-16 and KC-135 analyses.[24] The following parameters drive the design of the C-130 network:

- the TAI
- the geographic dispersion (for both home-station and contingency operating locations)
- the OPTEMPO for both home-station and deployed aircraft
- the extent of the economies of scale in the centralized workload
- the personnel costs ($65,000 per man-year)
- the aircraft shuttle cost ($5,286 per flying hour)[25]
- facility and equipment costs ($2 million per year per ISO inspection dock amortized).

The ISO inspection calendar will drive the majority of the demand and workload at the CRF. The TAI determines the total requirement for ISO inspection demand in the CRF network.[26] For home-station aircraft, the FY 2008 TAI beddown was used to determine the demand by operating location. Personnel and aircraft shuttle costs were determined as before using a C-130 block speed planning factor of 345 nm per hour. Personnel would cost $65,000 per manpower authorization, a C-130 cost per flying hour as indicated above.[27] We assumed that no training missions would be accomplished when shuttling the aircraft between the operating locations and the CRF, so the shuttle cost was determined to be 100 percent of the

---

[22] Fractional demands were not rounded to the nearest integer. If we considered the upper bound by rounding to the nearest integer, the pipeline requirement would be $42.4 million. This should be considered a limit because it does not consider the inventory requirement that was already allocated to the base for the base repair lead time or the rounding up of inventory associated with the depot pipeline.

[23] McGarvey, Carrillo, et al., 2009.

[24] See Appendix D and McGarvey, Carrillo, et al., 2009, for more details on these optimization models.

[25] Costs associated with shuttle to and from deployed FOLs were not included because current Air Force policy is to not permit ISO inspections to be performed in theater.

[26] The FOL off-equipment component repair and nongrounding failure workload creates sortie-generated workload that is considered additive to the network.

[27] Air Force Instruction 65-503, *US Air Force Cost and Planning Factors,* Annex 4-1, "BY 2008 Logistics Cost Factors," Washington, D.C.: Department of the Air Force, February 22, 2006.

total cost per flying hour.[28] We assumed that aircraft would be "clean" when sent to a FOL and would have a relatively long time to return before being sent to the CRF for an ISO inspection. Therefore, we did not measure the traditional operational effects of centralizing the C-130 fleet.

We analyzed the future operational requirements to determine the maintenance force necessary to support a variety of potential deployment requirements. We do not suggest that the Air Force build its maintenance manpower capability to any specific level. Rather, this is a force-sizing analysis built around a notional desired capability level. The analytic process that follows could be applied to other desired capability levels for the standard and specialty C-130 aircraft.[29] For the purposes of illustration, assume that the Air Force wishes to maintain a steady-state deployment of 40 percent of the standard CA C-130s and 60 percent of the specialty CC and CA C-130s. Again, while illustrative, these percentages are fairly consistent with OSD guidance.

Figure 3.7 shows the geographical distribution of the C-130 beddown for active-duty and AFRC units. Because of the long range of the C-130, we did not consider the USAFE and PACAF units separately from CONUS, but rather considered support to the C-130 global beddown. We assumed that the set of potential CRF sites is limited to the current C-130 operating locations and the three air logistics centers (ALCs), in Ogden, Utah; Oklahoma City,

Figure 3.7
FY 2008 C-130 Active-Duty and AFRC Worldwide Beddown

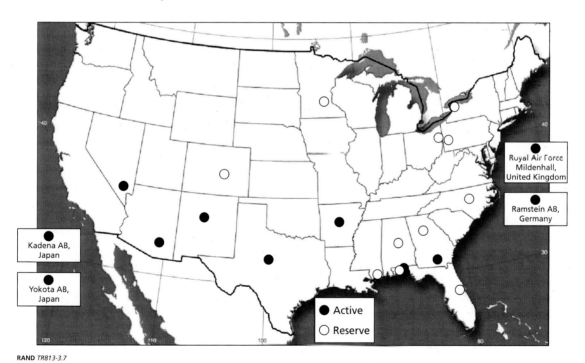

RAND *TR813-3.7*

---

[28] Our assumption is conservative in that we assume no operational training sorties can be aligned with shuttling an aircraft to and from a CRF. If sorties could be aligned with shuttling an aircraft, the transportation bill would be reduced.

[29] The OSD guidance discussed in Chapter One directs the services to plan and program for a future in which some fraction of the force is continuously deployed forward.

Oklahoma; and Warner Robins AFB, Ohio.[30] We further assumed that all CRF personnel work 40 hours a week. Because we assumed that there was no requirement for contingency-based CRFs, this analysis did not consider manpower dwell-to-deploy ratios or the fraction of forward-deployed manpower assigned to the active duty. These assumptions parallel our K-135 analysis and differ from our F-16 analysis.

The model considers every potential combination of CRF locations, from fully decentralized solutions to fully centralized solutions. However, because of the economies of scale that centralization offers and the higher utilization of facilities that two- and three-shift operations at CRFs provide, centralized network designs are more favorable than noncentralized designs.

For the combined active-duty and AFRC network, the optimization model identifies the minimum-cost solution as a single CRF at Little Rock AFB. Table 3.6 presents the performance of a set of CRF alternatives. Cost and manpower details are presented for the minimum-cost networks, as identified by the optimization model, for networks using various numbers of centralized facilities. Table 3.6 also shows the estimates of the average number of aircraft that would be at the CRF for repair or in transit between the CRF and the operating locations.[31] Increasing the number of CRFs in the network decreases the shuttling costs. However, networks with multiple CRF locations have decreased manpower staffing and shift requirements at each CRF. Consequently, the average number of days an aircraft is in the ISO inspection and/or refurbishment processes increases, and the increase in the number of aircraft in the ISO inspection process often more than offsets the decreased number of aircraft shutting between the CRF and the operating locations.

Figure 3.8 contrasts the performance of the best single-CRF (Little Rock AFB) with several other CRF solutions: one CRF at Robins AFB, two CRFs (Little Rock AFB and Yokota Air Base), a two-CRF solution (all standard C-130s at Robins, all specialty at Hurlburt Field), the optimal four-CRF solution, the optimal six-CRF solution, and the optimal ten-CRF solu-

Table 3.6
C-130 Active-Duty and AFRC CRF Network Options

	Number of CRFs				
	1	2	3	4	5
Annual costs ($M)					
Manpower	68.9	72.0	75.1	76.1	79.2
Shuttle	9.6	6.5	4.9	4.2	4.2
Facility	22.0	26.0	32.0	38.0	42.0
Total	100.5	104.5	112.0	118.3	125.4
Manpower positions (number)	1,059	1,107	1,155	1,173	1,221
Aircraft in ISO inspection, refurbishment, or transit (number)	12.7	14.2	15.7	18.7	19.2

---

[30] If C-130 units commonly traveled to cargo ports on the way to and from FOLs, the cargo ports could also be examined as possible candidates for centralization.

[31] We assumed transit times of four days round trip for aircraft shuttling between CONUS locations, four days round trip within PACAF and within USAFE, and seven days round trip between CONUS and PACAF or USAFE.

Figure 3.8
Comparative Cost of C-130 Network Designs

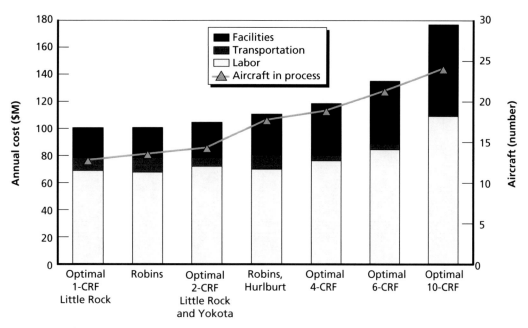

NOTE: Aircraft in process includes those in ISO inspection, refurbishment, or transit.

RAND TR813-3.8

tion.[32] In addition to the costs of each option, Figure 3.8 shows the estimated number of aircraft in the ISO inspection and/or refurbishment process or in transit (right vertical axis). Approximately $18 million separates the optimal one-CRF and the optimal four-CRF solutions, with several combinations of solutions displaying similar levels of performance in terms of cost and the number of aircraft in process or in transit. The insensitivity to location and number of CRFs allows a range of considerations beyond the scope of this analysis to figure into the final CRF location decision. For example, CRFs at Little Rock AFB and Hurlburt Field AFB provide a natural separation of standard and specialty aircraft; a CRF at Warner Robins might provide proximity to the depot; and possible permanent USAFE and PACAF CRFs might improve support to certain deployed forces.

Facility costs increase for each increasing CRF solution because, as the solutions increase the number of CRFs, the workload at each CRF decreases and so does the work-center staffing. As the number of shifts per day decreases at the CRFs, the average CRF flow time per ISO inspection and/or refurbishment increases at the CRFs. This means that additional CRF facilities are subsequently required to support the slower flow times at each CRF location. In Figure 3.8, a ten-CRF network generates the largest facility costs because fewer shifts worked at each of its relatively small CRFs.

The total steady-state manning requirement can be determined for any desired network. Suppose it was desired to implement the one-CRF solution at Little Rock AFB. The total

---

[32] The optimal four-CRF solution has CRFs located at Little Rock, Pittsburgh, RAF Mildenhall, and Yokota. The optimal six-CRF solution has CRFs located at Davis-Monthan, Little Rock, Moody, Pittsburgh, RAF Mildenhall, and Yokota. The optimal ten-CRF solution has CRFs located at Davis-Monthan, Dyess, Kadena, Keesler, Little Rock, Moody, Pittsburgh, Ramstein, Yokota, and Youngstown.

Figure 3.9
Total Cost Comparison of Current System and CRF Concept

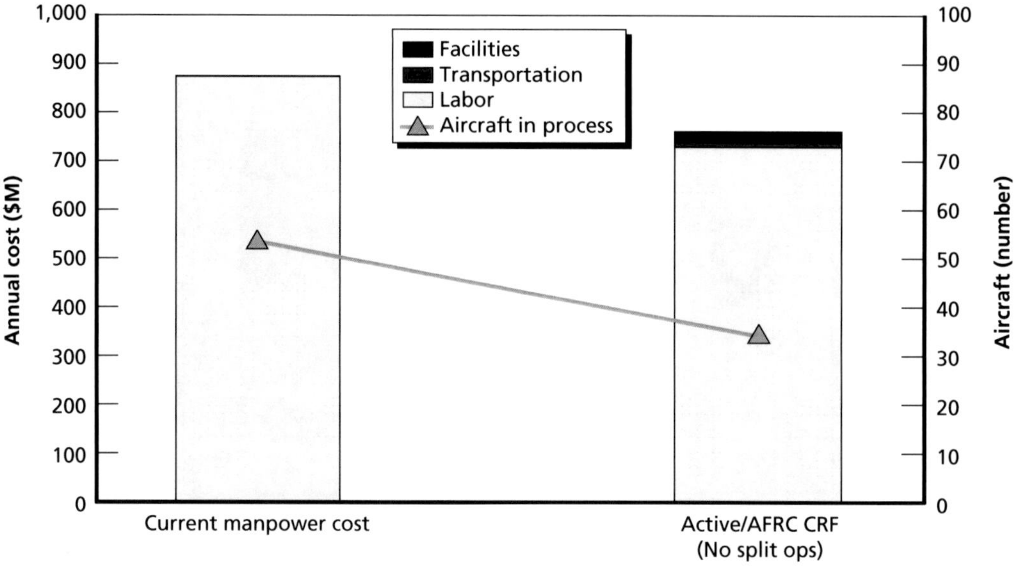

NOTE: Aircraft in process includes those in ISO inspection, refurbishment, or transit.

RAND *TR813-3.9*

steady-state manpower for this solution has 1,059 maintenance manpower positions at the CRF. If CRFs were placed outside CONUS, in PACAF and in USAFE, fewer than 100 maintenance manpower positions would need to be added to support the network of locations.

Total Manpower Requirements for the Mission-Generation and Network Facilities
Rebalancing the maintenance between the operating locations and the CRF network could free 2,516 maintenance manpower positions.[33] If the Air Force felt that the current force posture was adequate, this move could save an estimated $102 million annually. In addition, the centralization of the ISO inspection process would reduce the average number of aircraft in the ISO inspection by 19 aircraft a year. Figure 3.9 compares the costs and number of aircraft in the ISO inspection process in the current system with those of the proposed CRF concept.[34] The current system costs approximately $877 million while the total cost of the CRF network concept is less than $775 million for a net savings of $102 million.

By implementing the CRF concept, the Air Force could also use the manpower savings to improve the overall effectiveness of the current force. As with our F-16 and KC-135 analyses, we determined the manpower necessary to add a split-operations capability to each CC and CA squadron. We estimated that about 2,400 maintenance manpower positions would be necessary to fund split operations for the active-duty and AFRC network.[35] The $102 million

---

[33] Appendix A contains a detailed breakdown of authorizations.

[34] The analysis of the current system does not include the estimated average number of aircraft in the refurbishment process. Because this program is optional and is sometimes included just as part of the ISO inspection process, we did not determine the additional aircraft in this process. However, if the estimated number of aircraft in refurbishment was included, the estimated savings associated with centralization would increase.

[35] See Appendix A for manpower data.

cost savings from the CRF network could be applied to fund approximately 1,600 of the 2,400 positions without exceeding the current costs of the system.[36] Combining the CRF concept with fully funded split operations for the active-duty and AFRC network would cost approximately $26 million more than the current system. This additional cost could potentially be captured if the Air Force decided to implement the CRF concept for another weapon system and decided to apply the savings generated by networked maintenance for this other weapon system to C-130 maintenance.[37]

CRF Networks to Support the Total Force

An additional C-130 analysis assumed that the repair network construct supported active-duty, AFRC, and ANG units. We assumed that total force units receive both home-station and deployed support from the fixed CRF network. The analysis followed the steps as for the active-duty and AFRC network. Figure 3.10 represents the total force C-130 worldwide beddown.

First, we determined the baseline maintenance manpower.[38] The manpower rebalancing analysis did not address the 1,946 supervisory and support positions from the MXG and

Figure 3.10
FY 2008 C-130 Active, AFRC, and ANG Worldwide Beddown

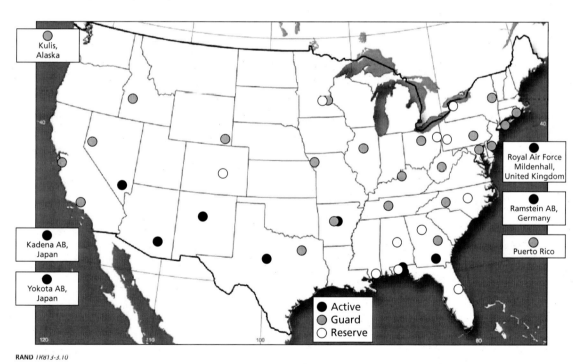

RAND IR813-3.10

---

[36] The costs of the 1,600 positions are assumed to be fully burdened active-duty maintainers. A mixture of active and AFRC full- and part-time personnel would increase the number of authorizations above 1,600.

[37] McGarvey, Carrillo, et al., 2009, identifies annual savings of $37 million and $43 million for the active-duty and AFRC F-16 and KC-135, respectively, if CRF networks were implemented with no split-operations plus-up.

[38] Appendix A offers a detailed discussion of this process.

Table 3.7
Manpower Requirements for the Total Force Mission-Generation Operations

	Current System			Total Force Network	
	ANG	Active Duty and AFRC			
		Standard	Specialty	Standard	Specialty
Group and MOS	621	800	525	1,334	612
AMXS FY08 UMD	1,440	3,612	2,960		
Mission generation concept					
Former AMXS (UTC-based)				4,740	2,771
Moved from MXS (UTC-based)				1,447	1,578
New total				6,187	4,349

NOTE: We again note that "Group and MOS" manpower values are not included in the calculations, and the "New total" only represents an increase to the AMXS or aircraft squadron.

MOS classifications across both aircraft category types (standard and specialty). However, as with the KC-135 analysis, we did include all maintenance manpower positions in the AMXS and the maintenance squadron (MXS) except AGE- and munitions-related maintenance.[39] We assumed that the 377 total force munitions authorizations remained constant and eliminated them from the remainder of this analysis and did the same with backshop-related JEIM and survival equipment–related authorizations.

We then determined a total force manpower requirement for the mission-generation work centers on the basis of individual squadrons, using the same approach as for active-duty and AFRC units. Table 3.7 summarizes the results. Across all units, the mission-generation work centers require 479 fewer AMXS positions than the current UMD count for C-130 AMXS. A further 3,003 authorizations are required to move from the MXS into the mission-generation work centers to create this mission-generation capability.

On completing the analysis for the mission-generation work centers, we again used the optimization model to determine the CRF network design for the total force. The optimization model identified the minimum-cost option as a single CRF at Standiford Field, Kentucky. Locating the CRF solution at Little Rock AFB instead would cost an estimated $125,000 more in annual transit than the Standiford Field solution, which is only about 0.1 percent of the total cost of the network solution.

We also considered what the optimal network design would look like if the network were forced to have multiple CRFs. As with the active-duty and AFRC analysis, we examined one-, two-, three-, four-, and five-CRF options. The one-CRF solution is comparable to the optimal solutions for two, three, and four CRFs. Table 3.8 presents the performance of a set of CRF alternatives. Cost, manpower, and aircraft in the ISO inspection process (including aircraft in transit to and from the CRF) are presented for a set of minimum-cost networks, each of which were identified by the optimization model as the cost-optimal network for a given number of CRFs. As with the active-duty AFRC CRF network, the solutions are somewhat insensitive to the number and location of the CRFs. For example, the optimal three-CRF network performs

---

[39] We assumed that the AGE manpower requirement at operational units did not change and that an additional AGE manpower requirement was necessary to support the CRF locations.

Table 3.8
C-130 Total Force CRF Network Options

	Number of CRFs				
	1	2	3	4	5
Annual costs ($M)					
Manpower	106.2	109.4	110.0	115.4	116.2
Shuttle	13.6	10.4	7.9	6.9	5.2
Facility	28.0	32.0	40.0	42.0	50.0
Total	147.8	151.8	157.9	164.3	171.4
Manpower positions (number)	1,633	1,682	1,692	1,774	1,775
Aircraft in ISO inspection, refurbishment, or transit (number)	15.3	18.0	22.2	21.1	24.6

nearly as well as the minimum-cost single-CRF network—the annual cost for three-CRF network is $10 million (7 percent) higher, and seven more aircraft are unavailable because they are in the ISO inspection process. These insensitivities suggest that, for one to four CRF network facilities, a broad range of other, more-qualitative considerations can influence the design of the CRF network but have little effect on cost and performance.

In addition to the mission-generation work centers and CRF manpower positions, we determined a further requirement for 682 FOL NGF and off-equipment manpower positions to support deployed operations.[40] Our analyses determined that implementing a CRF network concept for the C-130 would require 3,152 fewer manpower positions across the total force. This manpower reduction would generate total annual savings of approximately $103 million. In addition, centralization would mean that 35 fewer aircraft would be in the ISO inspection and refurbishment process at any time.

If, however, the Air Force wanted a split-operations capability for each C-130 squadron, it would require an additional 4,600 manpower positions, 1,600 of which could be funded by the $103 million cost savings. If the CRF concept were combined with fully funded split operations for the total force, it would cost approximately $120 million more than the current system cost.[41]

Considering the total force, the network freed an additional 636 positions to the 2,516 manpower positions from the active-duty and AFRC network. The relative contribution of ANG positions from the C-130 total force CRF network (636 of a total 3,152 authorizations) means that the C-130 total force network offers fewer manpower savings than the F-16 and KC-135 total force networks, for two reasons. First, the ratio of UTC requirements to UMD is relatively large for ANG C-130s, which suggests that current ANG C-130 manpower lies closer to its deployment requirement, thus allowing less potential for savings. Second, the additional CRF manpower necessary to support ANG C-130s is relatively large. This suggests that

---

[40] See Appendix A for a detailed breakout of the manpower figures.

[41] Robert S. Tripp, Ronald G. McGarvey, Ben D. Van Roo, James M. Masters, and Jerry M. Sollinger, *A Repair Network Concept for Air Force Maintenance: Conclusions from Analysis of C-130, F-16, and KC-135 Fleets*, Santa Monica, Calif.: RAND Corporation, MG-919-AF, 2010, extends the C-130 analyses to leverage existing facilities and infrastructure. In general, leveraging existing facilities could provide an additional savings of $10 million to 12 million.

C-130 CRF operations afford fewer manpower economies of scale per additional PAA added to the CRF network than do KC-135 CRF operations.

Including facilities and transportation costs, the total financial savings of a C-130 total force CRF network appear to be essentially the same as for the active-duty and AFRC network. However, because the ISO inspection process is traditionally a one-shift operation for the ANG, a large number of ANG aircraft are down for ISO inspection–related maintenance under the current structure. Implementing the CRF concept for the total force would almost double the number of aircraft that would no longer be in the ISO inspection process over the active-duty and AFRC CRF network, at an equal total system cost. These effectiveness gains are large, considering that ANG aircraft represent only one-third of the C-130 fleet.

AFSOC Centralized ISO Inspection Facility

AFSOC has implemented the CRF concept at Hurlburt Field to support the ISO inspection process for active-duty and AFRC AFSOC locations in CONUS and the 352nd Special Operations Group at RAF Mildenhall. AFSOC's primary motivation for moving to a centralized ISO inspection concept was to enhance AFSOC's deployment capabilities, reassigning the manpower associated with the ISO inspection process to other AMXS and MXS roles. In addition, as AFSOC aircraft have been highly tasked and are in high demand from warfighters, AFSOC wanted to reduce the amount of time its aircraft spent in the ISO inspection process.

AFSOC negotiated a contract with L-3 Communications to perform 75 ISO inspections annually across five C-130 variants for $12 million beginning in December 2007. The workload for each ISO inspection is negotiated 120 days before the induction into the ISO inspection process, and contractors work round-the-clock shifts on the aircraft. The contract was paid for with SOCOM funds, and the AFSOC maintainers who formerly conducted these aircraft inspections were reassigned to other maintenance roles.

We examined the performance of this centralized ISO inspection facility to assess the benefits AFSOC had observed and to compare the facilities performance against the estimated performance of the CRF concept. We began by isolating a subset of historic AFSOC ISO inspection fly-to-fly times.[42] We then repeated this process using 20 fly-to-fly times from the contractor ISO inspection facility. Figure 3.11 illustrates the difference in performance between the decentralized ISO inspection process and the centralized contractor ISO inspection process.[43] For the historic date, the average fly-to-fly time was 35 days, with a standard deviation of 22 days. The variation associated with this process can be seen in the probability distribution of the fly-to-fly days. Moreover, because some of the AFSOC units had fewer aircraft, the ISO inspection process was likely staffed with one-shift operations, which lengthens the fly-to-fly interval.

---

[42] We gathered REMIS data from January 2005 to November 2007 and calculated the average fly-to-fly time, the standard deviation of the fly-to-fly time, and the probability distribution of the fly-to-fly times.

[43] Probability distribution functions were fitted to the data samples to graphically represent the distributions of the fly-to-fly times. As of March 2008, the ISO inspection calendar shifted from 365 to 450 days. A limited number of other inspection tasks that were also on a 365-day interval were not extended to the 450 calendar. Because these other inspections were frequently performed during ISO inspections prior to March 2008, some of the observed differences in fly-to-fly times may be attributable to the additional inspections during the ISO process within the historic AFSOC data set.

Figure 3.11
Fly-to-Fly Comparison of Decentralized and Centralized ISO Inspections for AFSOC C-130s

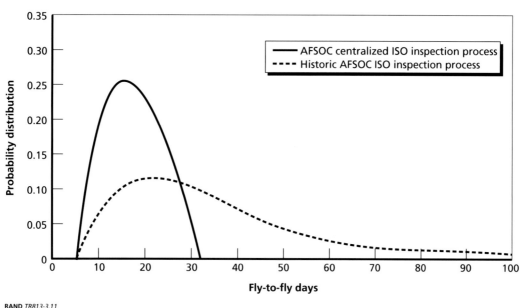

Centralization of the ISO inspection process provided two benefits for AFSOC. First, round-the-clock centralized operations reduced the average time aircraft are in the ISO inspection process. The average fly-to-fly interval decreased to 19 days, while the standard deviation decreased to nine days. This is reflected in the distribution function in Figure 3.11. Second, the variability of the ISO inspection process has decreased dramatically, improving scheduling for maintainers and operators.

In addition to comparing the performance of AFSOC's centralized ISO inspection process against the historic performance, we compared the cost and performance of the current, centralized system against the estimated performance of a centralized LCOM ISO inspection facility. The LCOM analysis was based on an AMC C-130 model and was not AFSOC-specific. Therefore, it was necessary to determine whether the AMC LCOM representation of the ISO inspection process was accurate for these other aircraft. Because RAND's CRF concept includes other workloads beyond ISO inspections, we isolated the ISO inspection process in the LCOM models and then used the simulation to determine manpower requirements and flow-time performance. Figure 3.12 summarizes the results. LCOM estimated that the manpower costs for supporting an ISO inspection facility for 75 aircraft using military personnel would be $11.4 million per year, which is similar to the centralized contractor cost of $12 million. LCOM estimated an average of 3.4 aircraft in the centralized ISO inspection process per year. This is also similar to the performance of the centralized process, which translates into an estimated 4.0 aircraft in ISO inspection at any time. These numbers validate the LCOM model estimates for specialty C-130s and our estimates of the benefits of centralization and the use of CRFs.

Figure 3.12
Comparison of LCOM-Estimated and Contractor Centralized ISO Inspection Facility

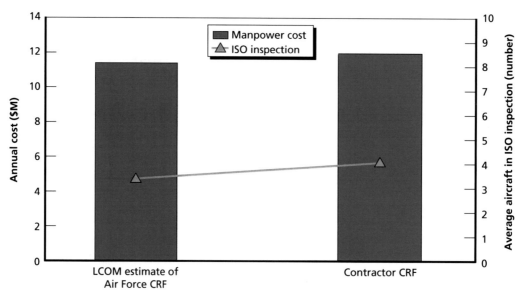

CHAPTER FOUR

Assessment of the Effects of Integrated Maintenance

The Air Force leadership has suggested that RAND extend beyond wing-level maintenance in future Logistics Enterprise Analysis phases to examine integrated maintenance concepts that unite wing-level and depot-level maintenance. One example of this idea is Warner Robins ALC's work on HVM, which could offer savings beyond those of the CRF concept by centralizing workloads and process improvements, such as by reducing redundant tasks seen in the current system. The CRF and HVM concepts should not be considered competing maintenance concepts. Rather, Air Force implementation of CRFs may be a logical stepping stone toward integrated maintenance concepts, such as HVM.

High-Velocity Maintenance

HVM integrates wing-level and depot-level workloads by incorporating them into depot maintenance processes. The goal of HMV is to increase the "touch" labor productivity rates of the depot maintenance worker and increase the flow rate of the PDM process. The initiative began in June 2007 with a cross-disciplinary team that examined current depot processes, including funding, requirements, infrastructure, materiel support, and information and technology. The HVM team examined the current PDM process and identified several factors that reduced the performance and capabilities of the ALCs, including a lack of understanding of the aircraft condition prior to induction; long, heavy maintenance cycles between PDMs; an inefficient and disorganized depot "job shop" production environment; and little or no planning and coordination of the supply process that supports PDM.[1] Combining these factors increased average flow time and variability and decreased labor productivity and aircraft availability.

The HVM concept was designed to address the several challenges the ALCs faced. The concept partitions the current PDM package into four smaller packages that rotate across four 18-month periods. The ISO inspection process would be folded into the reduced HVM PDM package. Each package would require approximately 4,500 hours, would have an average daily burn rate of 400 to 500 hours, and would be completed in approximately 13 calendar days. Across all four segments, each aircraft would spend a total of 52 days in the new HVM process over six years. The proposed process would greatly increase the depot's knowledge about the condition of the aircraft. Some of the workload for future visits could be preplanned, which would improve the planning process for both depot capacity and part supplies. In addition, a detailed process flow could be tailored to meet the requirements of each aircraft before it arrives at the ALC.

---

[1] The Warner Robins HVM team supplied this background information on HVM during site visits and in a briefing.

Significant efforts, funding, and sponsorship have gone into the HVM process thus far. HVM team members span several organizations, including Tinker ALC and Ogden ALC, multiple MAJCOMs, the Defense Logistics Agency, and the Global Logistics Support Center. Warner Robins has applied for and received $16.7 million in Air Force Smart Operations 21 funding to review, develop, and sequence processes and to purchase equipment and technical data. The positive gains the HVM team has cited include a significant waste reduction that may reduce existing costs of PDMs by 10 to 25 percent. A pilot project to validate HVM concepts on AFSOC C-130 aircraft is targeted for FY 2009 and will increase the number of aircraft moving through the process in the coming years. The team hopes to extend the processes and capabilities developed under HVM for the C-130 to other weapon systems and ALCs.

Current Depot-Level Performance

Although HVM is still in the proof-of-concept stage, it is possible to estimate the general effect of the integrated maintenance concept by combining empirical data for the existing PDM process with the planned HVM parameters for the assessment. We used PDM data from the Air Force Total Ownership Cost–Cost Analysis Improvement Group for FY 2005 through July 2008 to estimate average PDM costs and associated labor hours. The sample was 210 C-130 PDMs: 96 from Ogden ALC and 114 from Warner Robins ALC. Of the 210 PDMs, 136 were conducted on C-130Es or C-130Hs, and the rest were conducted on specialty variants of the C-130. The FY 2008 wage grade cost, $65,800 a year, was used as the average salary for a civilian maintainer.[2] The productivity of the maintainer was an estimated 49.64 percent.[3]

The calculated mean number of charged depot MMH for PDMs was 16,722 hours, with an average of an additional 60 hours for modifications beyond the planned PDM package. The standard deviation of MMH for the PDM process was 3,296 hours. Figure 4.1 is a histogram of the associated MMH of the PDM process (including modifications) charged to the Air Force for standard and specialty aircraft. Differentiating between standard and specialty aircraft, the data sample revealed that specialty variants were more likely to have additional work done in conjunction with the PDM than the standard variants; 58 percent of the specialty aircraft had analytic condition inspections or ISO inspections or other special inspections conducted on the aircraft in conjunction with the PDM, compared to 34 percent of the standard aircraft. These additional tasks likely contributed to the difference in the average MMH of PDM for standard aircraft (16,077) compared with specialty aircraft (18,560). In addition, it shows that AFSOC and the other MAJCOMs are already integrating the PDM and ISO inspection workloads when possible.

In addition to the financial information from Air Force Total Ownership Cost–Cost Analysis Improvement Group, we gathered PDM production data (FY 2007–July 2008) from Warner Robins ALC to estimate flow times of the PDM process and the effects of ISO inspections on the PDM flow time. The sample data were organized by whether an ISO inspection was conducted during the PDM and the type of ISO inspection (major or minor) that was conducted. Table 4.1 summarizes the data. The average PDM flow time across all samples was

---

2   Air Force Instruction 65-503, 2006.

3   Warner Robins ALC productivity rates from FY 2003 through July 2008.

Figure 4.1
Histogram of the Distribution of PDM Man-Hours, FY 2005–July 2008

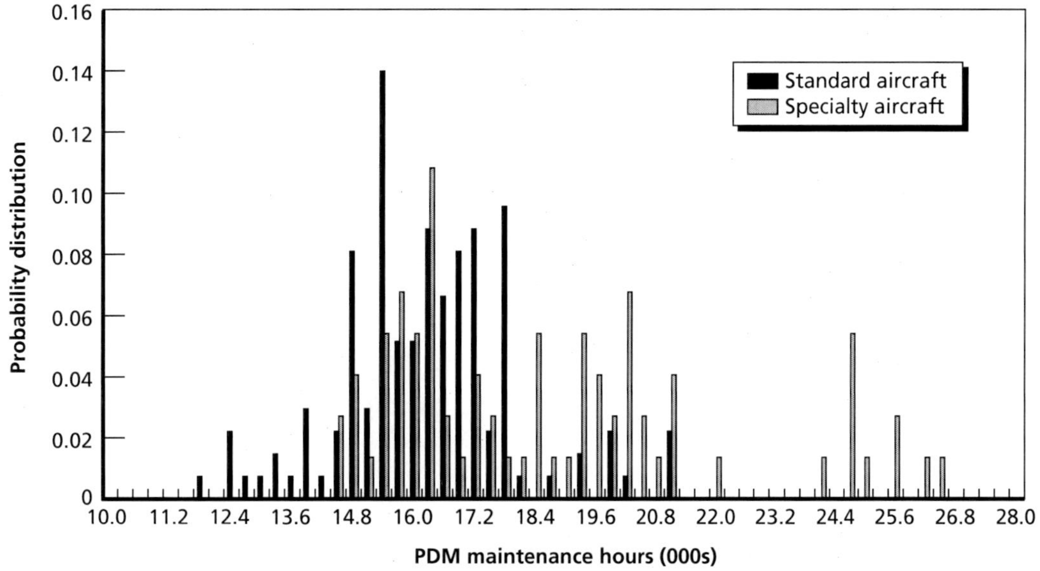

RAND TR813-4.1

Table 4.1
Warner Robins C-130 PDM Process Flow Data: FY 2007–July 2008

MAJCOM	No Isochronal Inspection		Minor Isochronal Inspections		Major Isochronal Inspections		Total Average	
	Number	Average Flow Days	Number	Average Flow Days	Number	Average Flow Days	Number	Average Flow Days
AETC	1	184	3	206	—	—	4	201
USAFE	5	312	—	—	—	—	5	312
AFRC	20	176	—	—	—	—	20	176
ANG	16	167	2	194	—	—	18	170
AFSOC	9	192	6	199	3	244	18	203
Total	51	189	11	200	3	244	65	196

NOTE: Totals may not sum exactly due to rounding.

196 calendar days, and the standard deviation of the PDM process was 64 calendar days. Table 4.1 suggests that ISO inspections increase the overall flow time for the PDM process. However, the significant variation across the relatively small PDM data sample makes the true effect of ISO inspections on the PDM flow rate unclear.

Preliminary Analysis of High-Velocity Maintenance

Having observed that rebalancing workloads between operating locations and CRFs for wing-level workloads led to economies of scale and reduced manpower requirements, we extended the analysis by incorporating the depot-level workload. We compared the performance parameters for the existing PDM and CRF processes to those estimated for the HVM process to determine whether HVM offered savings beyond those of the CRF concept still employing the existing PDM process.

We assumed that the maintenance and flow-time estimates Warner Robins provided would capture the current PDM-related workload and the ISO inspection–related and refurbishment-related workload in the CRF process. We did not assume that the depot would handle centralized off-equipment component repair but did assume that some centralized facility would handle it.[4] In the existing PDM process, approximately 60 aircraft per year undergo a PDM. Although all aircraft undergo the ISO inspection process, not all aircraft undergo the PDM process (such as newer aircraft). To account for the ISO inspection cycle, we assumed that all aircraft in the TAI would undergo the HVM process. This meant adjusting the assumed demand for HVMs to match the demand requirement for the TAI. We estimated approximately 95 PDMs per year for a direct comparison of the estimated workload and costs of each maintenance concept. Using the previously mentioned estimated MMH, labor costs, and productivity factors, we calculated the total estimated costs of the current PDM process. In addition, we calculated the cost of shuttling aircraft between depots and operating locations.

We developed a comparison model using Warner Robins estimates to determine the cost and performance parameters of the HVM process. Using it, we changed the schedule of the demand for the C-130 TAI and matched it to an 18-month calendar, used the MMH estimates and combined that value with the standard labor rate and productivity rates, and calculated the shuttling costs of sending aircraft between the operating locations and the depot. We assumed that an HVM facility would incur the same facilities costs as would a new CRF facility. Although existing depots may already have sufficient capacity to implement HVM without building additional infrastructure because they can leverage existing depot capacity, this capacity would not necessarily be available for the C-130 HVM process.[5]

Table 4.2 is a summary comparison of the HVM and CRF concepts. HVM offers more savings in efficiencies and effectiveness for both the active-duty and AFRC network and the total force network than CRF alone. Because the HVM process relies on higher levels of hands-on labor than the existing PDM process, using HVM would free approximately 33 active-duty and AFRC aircraft from the PDM/ISO inspection process, in addition to those the CRF network already frees.[6] If the Air Force felt that the current posture was adequate for the active-duty and AFRC and total force solutions, implementing HVM could save an additional annual savings of $30 million and 73 million, respectively. If the Air Force implemented HVM as a means of increasing effectiveness, the additional savings could be applied to split-operations manpower plus-ups. We estimate that an additional 400 positions could be funded in the active-duty and AFRC network, and an additional 1,100 positions could be applied to the total force network.[7]

---

[4] The centralized off-equipment repair captures economies of scale in the maintenance process. The centralized workload could be performed at a depot facility, but no information was available to estimate the potential savings associated with civilians performing this workload.

[5] We did not assume that materials costs would differ from our calculations of the inventory pipeline estimates discussed in Chapter Three.

[6] We consider transit delays for both the CRF network and the HVM network.

[7] These positions are assumed to be fully burdened active-duty maintenance personnel. If the positions were a mix of active-duty, ANG, and AFRC full- and part-time positions, additional split-operations positions could be funded.

Table 4.2
Comparison of CRF Concept and HVM Concept for the Active-Duty and AFRC Network

Summary of Analysis	CRF Network with Existing PDM Process	HVM Concept
Active-duty–AFRC network		
Fewer Aircraft in PDM/ISO	19	52
Option A: savings with current posture ($)	102	132
Option B: split-operations positions funded (number)	1,600	2,000
Total force network		
Fewer Aircraft in PDM/ISO	35	86
Option A: savings with current posture ($)	103	176
Option B: split-operations positions funded (number)	1,600	2,700

Figure 4.2 illustrates the difference in the total costs and the number of aircraft in the PDM/ISO inspection process.[8] As with the CRF analysis, we estimated the change in the number of aircraft that would be in the PDM or the ISO inspection process in each scenario.[9] Approximately 33 fewer aircraft would be tied up in the PDM process annually in the active-duty and AFRC network; 51 fewer in the total force network.

Figure 4.2
Comparison of CRF and HVM Concept for the Active-Duty and AFRC Network

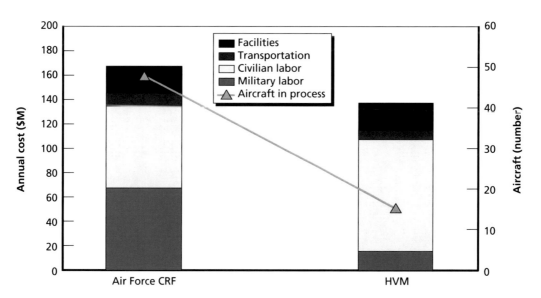

NOTE: Aircraft in process includes those in PDM, ISO inspection, refurbishment, or transit.

RAND TR813-4.2

---

8   The HVM process would also handle these refurbishment-related tasks in the 18,000-hour, four-stage cycle.

9   The estimated aircraft in refurbishment process was somewhat difficult to determine, but there should be aircraft in addition to our estimate of 19 per year.

CHAPTER FIVE

Conclusions

The U.S. Air Force has identified the need to realign its logistics enterprise with the changes in the national security environment. The analysis in this report has examined alternatives that support the Air Force's objective by rebalancing the resources invested in mission-generation maintenance and network-supported maintenance at C-130 active-duty and AFRC locations, along with the implications of extending these concepts to the Total Force. The report also provided an initial examination of integrating wing- and depot-level maintenance at centralized facilities, using such concepts as HVM.

The Air Force can maintain its C-130 fleet using significantly fewer resources or can increase operational unit maintenance capabilities at a cost comparable to that of the current system by reallocating maintenance resources from unit backshops to a centralized network. Centralizing specific maintenance workloads achieves economies of scale for maintenance resources across the fleet. These financial benefits are slightly offset by the costs for creating centralized facilities and shuttling aircraft and components to and from centralized facilities and the manpower diseconomies associated with splitting workloads between operating locations and centralized facilities.

Centralizing only active-duty and AFRC maintenance would save about 2,500 authorizations and yield an annual savings of $102 million, after accounting for increased expenditures due to CRF construction and transportation of aircraft and components between operating locations and CRFs. Including the ANG would increase the personnel savings to 3,200 and yield an annual savings of about $103 million. The Air Force could apply these savings toward funding 1,600 split-operations positions in the active-duty and AFRC network, and 2,000 in the total force network.

If the Air Force felt that current capabilities were sufficient, allocating maintenance activities to CRFs would yield significant financial savings. These savings could be applied to other areas beyond aircraft maintenance.

In addition to providing enhanced maintenance capabilities, by implementing the CRF concept, **the Air Force would realize gains in operational effectiveness of the C-130 fleet.** CRFs would employ two- and three-shift operations in most work centers. HVM would also employ multiple shifts with more personnel working on each aircraft. The increased aircraft flow rate through the maintenance process would decrease the number of aircraft that are tied up in the system and therefore unavailable to the operator. The Air Force could realize an increase of approximately 5 to 10 percent of the C-130 TAI released from scheduled maintenance activities.[1]

A number of potential network configurations with alternative potential CRF locations have comparable total system costs. This feature provides a great deal of discretion in selecting potential sites for CRFs within the vicinity of the optimal solution. Air Force leadership can use the range of solutions provided in this research to weigh the design alternatives and external factors when choosing the desired capabilities level, investment level, and network design.

---

[1] Note that, while fewer aircraft would be in the aircraft inspection process at any time, not all of them would be expected to be available because some would likely be unavailable for other reasons, such as engine failure.

Maintenance Manpower Authorizations

Determining C-130 Maintenance Manpower Authorizations

C-130 wing-level maintenance manpower levels were determined using MPES data, which were refined using the procedure discussed in this appendix.

RAND's source for UMD manpower authorizations data was the end-of-month data extract from MPES, the data management system Headquarters Air Force, Manpower and Personnel, Integration Division (AF/A1MZ) uses to collect, manage, and consolidate data on Air Force manpower requirements and funded authorizations. Each end-of-month extract that RAND receives from this office represents a consolidation, as of that point in time, of UMDs for all Air Force (including AFRC and ANG) organizations and locations into a single data table.

C-130 wing-level maintenance is currently organized under an MXG comprising four squadrons: AMXS, CMS, EMS, and the MOS. Personnel assigned to AMXS, CMS, and EMS conduct the hands-on maintenance tasks. These squadrons are further divided into work centers or "shops," according to the maintenance tasks that they perform. Table A.1 presents the set of work centers for a typical C-130 MXG.

The MPES data contain all manpower authorizations, including maintenance manpower. To segregate the manpower records of interest from the entire MPES data set, the Air Force program element (AFPE) codes and titles were first used to identify all C-130–related authorizations.[1]

We then identified the subset of these authorizations associated with maintenance by examining two additional MPES fields: the functional account code (FAC) and organizational title (ORGT). We classified maintenance manpower positions into three categories:

1. **MXG**, consisting of all records with an ORGT indicating maintenance operations, plus any additional records with FAC equal to 21*** (that is, with leading digits 21) whose ORGT is not equal to aircraft maintenance.
2. **AMXS**, consisting of all records with an ORGT indicating Aircraft Maintenance, plus any additional records with FAC equal to 22*** whose ORGT is not equal to maintenance operations.
3. **EMS and CMS**, consisting of all other records with ORGT indicating maintenance, component maintenance, equipment maintenance, munitions, plus all other records with FAC equal to 23***.

---

[1] We used the following set of AFPEs in this analysis: 41115, 41897, 27253, 27224, B8048, 41132, 27597 (only considered at Little Rock AFB), 54332, 53114, 54314, 54343, 53124, and 53122.

Table A.1
C-130 Maintenance Work Centers

Aircraft MX Squadron	Component MX Squadron	Equipment MX Squadron
Crew chiefs	Propulsion flight	Aero repair
Specialists	Avionics test stations	Structural repair
Flightline propulsion	Pneudraulics	Metals technology
Flightline E&E	Fuels	AGE flight
Flightline communications and navigation	E&E	Munitions flight
Flightline GAC		Survival equipment
Flightline ECM		

The AFRC and ANG data, however, presented an additional complication, because some records reflected full-time personnel authorizations while others reflected part time. To differentiate between full- and part-time personnel, the resource identification code and title were used, which identifies personnel type. This field typically categorized each record by one of four main groups: active-duty officers and enlisted personnel, drill officers and airmen, guard and reserve technicians, and nontechnician civilians. The following logic was used to calculate the full- versus part-time authorizations:

Full time = Total active-duty enlisted + Total guard and reserve technicians
+ Total nontechnician civilians

Part time = Total drill officers and airmen – Total guard and reserve technicians.

Units that are assigned both the C-130 and another MDS presented another complication because it can be difficult to distinguish the manpower positions that support the C-130 from those that support the other MDS. In many, but not all, instances, the AFPE filter helped isolate C-130 positions. For example, AFPE 27224M, supporting "Combat Rescue & Recovery Maintenance," appears in records associated with Moody AFB. Because both the HC-130P and HH-60 are assigned to Moody and both are associated with this single AFPE, it is difficult to distinguish between the C-130 and non–C-130 positions. To address this problem, we eliminated the positions whose Air Force Specialty Codes clearly supported helicopter maintenance and included all other maintenance positions associated with AFPE 27224M at Moody AFB.[2] However, because such uncertainties are limited to very few of the AFPEs included in this analysis, these overestimates are likely rather small.

Associate units also counted in totaling C-130 manpower authorizations. Under the Total Force Integration (TFI) concept, there are three types of associate arrangements: classic associate, active associate, and Air Reserve Component associate. In an associate arrangement, a cadre of personnel from one unit is permanently assigned to associate (work) with a unit of another component (the principle component) at that unit's location.[3] Table A.2 identifies the Air Force component with principal responsibility for the MDS and the partner component for each arrangement.

---

[2] Note that this introduces a slight overestimate to the count of C-130 maintenance manpower.

[3] John G. Drew, Kristin F. Lynch, James M. Masters, Robert S. Tripp, and Charles Robert Roll, Jr., *Options for Meeting the Maintenance Demands of Active Associate Flying Units*, Santa Monica, Calif.: RAND Corporation, MG-611-AF, 2008.

This analysis of C-130 maintenance manpower levels counted the authorizations of the units involved in TFI initiatives according to the MAJCOM they support. For example, the active-duty maintenance personnel of the 30th Airlift Squadron at Cheyenne Regional Airport support an ANG unit. For this reason, the maintenance manpower authorizations within the 30th Airlift Squadron were included in the total ANG authorizations. Further detail on the TFI instructions that affected our results can be found at the end of this appendix.

A final consideration in calculating the C-130 maintenance manpower authorizations was whether the authorization pertained to servicing standard C-130s or specialty MDS (e.g., MC-130, HC-130, and WC-130). For some locations that operate both standard and specialty C-130s (e.g., Keesler AFRC), while we can be confident that the total C-130 maintenance manpower counts are accurate, it was sometimes difficult to split these data accurately between standard and specialty manpower. Once the distinction was made between full- and part-time positions for the reserve and guard data and once all multiple-MDS and associate unit issues were addressed (to the best of our ability), the authorizations were separated into standard and specialty authorizations to obtain the counts presented in Tables A.3 and A.4.

Table A.2
Associate Unit Arrangements

Arrangement Type	Principal Component	Partner Component
Classic associate	Active duty	AFRC/ANG
Active associate	ANG	Active duty
ARC associate	AFRC/ANG	AFRC/ANG

Table A.3
Standard C-130 Maintenance Personnel Authorization Totals

Organization	Active duty	AFRC		Active duty and AFRC	ANG		Total
		FT	PT		FT	PT	
Aircraft MXS	2,418	288	906	3,612	462	703	4,777
Equip. & Comp. MXS	1,539	606	569	2,714	1,251	1,659	5,624
MXG & Ops	423	237	140	800	290	244	1,334
Total	4,380	1,131	1,615	7,126	2,003	2,606	11,735

Table A.4
Variant C-130 Maintenance Personnel Authorization Totals

Organization	Active duty	AFRC		Active duty and AFRC	ANG		Total
		FT	PT		FT	PT	
Aircraft MXS	2,663	87	210	2,960	123	152	3,235
Equip. & Comp. MXS	2,126	261	340	2,727	290	299	3,316
MXG & Ops	400	71	54	525	50	37	612
TOTAL	5,189	419	604	6,212	463	488	7,163

Associate Arrangements Considered

Using documentation available on the TFI web page, we identified project-specific associate arrangements and applied them to the MPES data. The following highlight the relevant TFI associate unit instructions and discuss the affected units and MPES data:

- **Establish a BRAC-directed active-associate unit on ANG C-130s at Cheyenne Municipal Airport (Office of Primary Responsibility: ANG; Office of Collateral Responsibility: AMC).** MPES data were found to support this instruction. Multiple sources identified the 30th Airlift Squadron as the active component of an active-associate unit.[4] Since the active-duty personnel support the ANG, the maintenance authorizations within the 30th Airlift Wing were identified and realigned to the ANG totals.
- **Establish a BRAC-directed active-associate unit on AFRC C-130s at Pope AFB (Office of Primary Responsibility: AFRC; Office of Collateral Responsibility: AMC).** MPES data was found to support this instruction. The 440th Airlift Wing fact sheet identifies the 2nd Airlift Squadron and the 43rd Aeromedical Evacuation Squadron as active-associate units within the 440th Airlift Wing's operations group. For this reason, the maintenance authorizations within the 2nd Airlift Squadron and the 43rd Aeromedical Evacuation Squadron were totaled and realigned to the AFRC totals.

Total Manpower Requirements for the Mission-Generation and Network Facilities

To determine the total maintenance manpower requirement across the rebalanced enterprise of mission-generation and network-maintenance facilities, we summed the two sets of requirements in Table A.5 and compared the current FY 2008 UMD for active-duty and AFRC to the rebalanced CRF network alternative. The UTC-based approach for determining the mission-generation work centers required 1,399 fewer AMXS positions than the current UMD count for C-130 AMXS. A further 2,017 authorizations are moved from the MXS into the mission-generation work centers to create this mission-generation capability. The remaining workload that was not rebalanced from the MXS to the mission-generation unit would be sent to the CRF, which requires 1,059 positions. The network requires an additional 562 positions to address the FOL-generated NGF and off-equipment workload.[5] Table A.6 includes the split-ops requirements for the active-duty and AFRC network, and Tables A.7 compares the FY 2008 UMD for the total force, the rebalanced total force CRF network alternative, and the rebalanced total force network with split-ops plus-ups.

---

[4] Wyoming National Guard Public Affairs, "30th Airlift Squadron settles in at Wyoming Air National Guard," Cheyenne: Wyoming Air National Guard 153rd Airlift Wing, December 4, 2006.

[5] The FOL and NGF workload is a function of the assumed 40-percent standard and 60-percent specialty deployment aircraft.

Table A.5
Manpower Requirements for Active-Duty and AFRC C-130 Network

| | Current System | | | Active-Duty and AFRC CRF Network | |
| | | Active-Duty and AFRC | | | |
	ANG	Standard	Specialty	Standard	Specialty
Group and MOS	621	800	525	800	525
AMXS FY08 UMD	1,440	3,612	2,960		
Mission generation concept					
Former AMXS (UTC-based)				2,708	2,443
Moved from MXS (UTC-based)				859	1,180
New total				3,567	3,623
MXS FY08 UMD	3,236	2,431	2,335		
CRF concept					
CRF Network				—1,059—	
FOL NGF Workload				350	212
New total				—1,621—	
Total	4,676	—11,338—		—8,811—	

Table A.6
Manpower Requirements for Active-Duty and AFRC C-130 Network Supporting Split Operations

| | Current System | | | Active-Duty and AFRC CRF Network | |
| | | Active-Duty and AFRC | | | |
	ANG	Standard	Specialty	Standard	Specialty
Group and MOS	621	800	525	800	525
AMXS FY08 UMD	1,440	3,612	2,960		
Mission generation concept					
Former AMXS (UTC-based)				2,708	2,443
Moved from MXS (UTC-based)				859	1,180
Additional split-ops plus-up				1,642	753
New total				5,209	4,376
MXS FY08 UMD	3,236	2,431	2,335		
CRF concept					
CRF Network				—1,059—	
FOL NGF Workload				350	212
New total				—1,621—	
Total	4,676	—11,338—		—11,206—	

Table A.7
Manpower Requirements for the Total Force C-130 Network

| | Current System | | | Total Force Network | |
| | | Active Duty and AFRC | | | |
	ANG	Standard	Specialty	Standard	Specialty
Group and MOS	621	800	525	1,334	612
AMXS FY08 UMD	1,440	3,612	2,960		
Mission generation concept					
Former AMXS (UTC-based)				4,740	2,771
Moved from MXS (UTC-based)				1,447	1,578
Additional split-ops plus-up				3,517	1,113
New total				9,704	5,462
MXS FY08 UMD	3,236	2,431	2,335		
CRF concept					
CRF network				—1,633—	
FOL NGF workload				458	224
New total				—2,315—	
Total		—16,014—		—17,481—	

Analysis of ISO Inspections and HSCs Using REMIS

Our analysis built on previous work that examined the REMIS data for both the F-16 and KC-135.[1] As with the previous study, we chose to gather and analyze empirical data as recorded in the Air Force's REMIS maintenance data collection system. With the assistance of the REMIS Office, we collected five years (January 2002 to January 2007) of maintenance data for the C-130 fleet. The C-130 data sample contained a total of 8,972,647 MMH.

As with the previous study, our methods and objectives in this analysis contained several components. First, we conducted a basic assessment of REMIS as a tool and its capabilities in properly capturing the work that occurs during ISO inspection and HSC maintenance periods. Second, we built on our previous analysis of ISO inspection and HSC inspection periods by calculating three important metrics associated with the fly-to-fly maintenance periods for the ISO inspection and HSC processes, including the fly-to-fly time; the ISO-HSC on-aircraft flow time, which is defined as the estimated number of consecutive calendar hours throughout the ISO inspection or HSC in which on-equipment maintenance was conducted on the aircraft; and the MMH of a fly-to-fly ISO inspection. Third, we analyzed attributes of ISO inspection and HSC inspections across C-130 variants and MAJCOMs.

ISO inspections are categorized as major or minor, depending on the work unit code. We chose to categorize them all as ISO inspections. This is due in part to the fact that we are considering an infinite time horizon. If we were looking at a specific period, we would categorize them separately. We understand there is a difference between means and standard deviations of major and minor ISO inspections and that, at the operational level, the unit wants to know what type of ISO inspection is coming in.

To determine these values, we analyzed REMIS on-equipment maintenance records by each aircraft serial number and the date and time of maintenance action. We used the five-digit work unit code to identify ISO inspection and HSC events. We then sorted aircraft sortie records by aircraft serial number and the date and time of the sortie. We then combined the maintenance and sortie records by each aircraft and date and time. After we categorized all maintenance records into fly-to-fly ISO inspection and HSC maintenance and other maintenance periods, we estimated the theoretical flow time for the ISO inspection and HSC by tagging every hour of each day in which on-equipment maintenance occurred. We then removed all the hours in which no maintenance occurred (nights, weekends, etc.) to come up with the on-aircraft flow time. Finally, we determined the total MMH by summing all the maintenance actions that fell within the fly-to-fly period.

---

[1]   For a complete description of the methods used and the motivation behind using these methods see McGarvey, Carrillo, et al., 2009.

Error-Reduction Process

This section explains the manual process for reducing errors in the REMIS ISO inspection data and again for the HSC data. The initial query to REMIS used work unit codes to identify ISO inspection and returned 736,649 instances of possible events. We organized these by aircraft serial number and the start and end times of each maintenance period.

On visual inspection of the observations, we discovered that the query had returned some instances in which no ISO inspection occurred. To eliminate such spurious data, we created a macro in Microsoft Excel that identified observations for which an ISO inspection had occurred and the aircraft sorties that occurred immediately before and after that ISO inspection, then eliminated all other observations. We also identified and eliminated observations with no start or end dates, which occurred when the start or end dates fell outside our data range. This reduced the number of observations to 5,201.

The next major step was to examine the remaining observations to identify any data problems that still existed. We first marked any occurrences of consecutive dates of ISO inspections occurring with a sortie in between. For example, one C-130 had an ISO inspection end on January 22, 2006, then flew a sortie on the same date, and then entered into another ISO inspection on January 23, 2006. There were also instances where the ISO inspection ended on January 22, 2006, the aircraft flew a sortie and then entered into another ISO inspection on the same date. We also marked occurrences of the same aircraft having duplicate ISO inspections listed under different units. For example, an MC-130 with the 16th Special Operations Wing had an ISO inspection from July 23, 2004, to August 19, 2004. This same aircraft was also listed under the 9th Special Operations Squadron as having had an ISO inspection during the same period.

The final step was to take consecutive dates of ISO inspections occurring with a sortie in between, along with occurrences of ISO inspections for a single aircraft listed under different units, and consolidate these data into one observation. Instead of creating another macro, we chose to do this manually to ensure that no remaining observations were eliminated. The consolidated data resulted in 3,728 ISO inspection observations.

We used a similar process to identify HSCs, which resulted in a consolidated data set containing 3,836 HSC observations. For verification, we compared maintenance data from several C-130 units listing when their aircraft were in ISO inspections and HSCs with our consolidated REMIS data, and found that the lists matched up. Where discrepancies existed, we consulted with the units and adjusted the data accordingly.

Analysis of REMIS Data

We used the REMIS data to examine the fly-to-fly times, the on-aircraft flow times, and MMH for ISO inspections and HSCs. First, we determined the fly-to-fly ISO inspection and HSC times, which we define as the number of days between the last sortie that precedes the ISO inspection or HSC and the first sortie that follows the ISO inspection or HSC. Second, we estimated the on-aircraft flow time. Third, we calculated all on-equipment MMH for the fly-to-fly ISO inspection or HSC event.

Tables B.1 and B.2 summarize the results for both the HSCs and ISO inspections. Both tables are organized by MAJCOM. The average and standard deviation were calculated for each of the three metrics.

Table B.1
C-130 ISO Inspection Fly-to-Fly Times, On-Aircraft Flow Times, and Maintenance Man-Hours, by MAJCOM

MAJCOM	Fly-to-Fly Times (days)		On-Aircraft Flow Time (days)		Maintenance Man-Hours	
	Average	Std. Dev.	Average	Std. Dev.	Average	Std. Dev.
ACC	39.29	15.99	17.42	6.15	3,686	893
AETC	32.08	14.36	14.06	5.33	2,245	978
USAFE	34.66	12.26	15.38	4.73	2,333	812
AFRC	50.30	19.92	16.24	6.34	2,489	1133
AMC	36.09	17.06	14.97	5.70	2,464	946
ANG	63.13	19.79	16.63	5.45	2,402	907
PACAF	34.32	15.53	12.32	4.05	1,517	577
AFSOC	26.55	14.87	12.79	8.36	2,633	1,566

Table B.2
C-130 HSC Fly-to-Fly Times, On-Aircraft Flow Times, and Maintenance Man-Hours, by MAJCOM

MAJCOM	Fly-to-Fly Times (days)		On-Aircraft Flow Time (days)		Maintenance Man-Hours	
	Average	Std. Dev.	Average	Std. Dev.	Average	Std. Dev.
ACC	15.10	8.30	4.79	2.65	402	231
AETC	13.76	6.88	4.47	2.24	338	219
USAFE	14.72	9.06	5.25	2.52	508	250
AFRC	15.51	9.47	3.64	2.17	326	220
AMC	13.49	9.49	4.35	2.73	319	223
ANG	14.14	9.84	2.71	1.91	255	197
PACAF	12.90	7.96	3.96	2.20	375	241
AFSOC	12.77	8.07	4.82	2.87	473	303

Modeling C-130 Maintenance with the Logistics Composite Model

LCOM is a stochastic discrete-event simulation model that is most commonly used to determine the manpower requirements associated with direct aircraft maintenance activities. These tasks involve preparation of aircraft on the flightline, repair of aircraft and aircraft components that fail during flight operations, and the upkeep of airframes during any scheduled maintenance activities.

Fundamentals of LCOM

Within the simulation, the analyst may regulate the availability of any of four key resources. While manpower levels are of primary importance, the user also has control over the amount of maintenance equipment, spare parts, and repair facilities on which the simulation may draw. The interplay between these four fundamental resources defines LCOM's "composite" nature.

The key driver in LCOM involves satisfying the demand for sorties in a preprogrammed flight schedule. When a sortie is required, an aircraft is selected from the pool of available airframes. Once the sortie has been flown, LCOM determines whether any parts or aircraft subsystems require maintenance. If a repair is deemed necessary, LCOM simulates the requisite repair networks that will ensure the aircraft can be returned to flight operations.

LCOM first handles maintenance requests by verifying that the resources needed for the repair are available. If LCOM determines that resources are insufficient (for example, no spares are currently on hand, or a specialist needed for the repair has already been allocated to a different task), one of two outcomes is possible. One option the simulation may take is to defer the maintenance action until any necessary resources become available. The alternative is to reallocate the needed resources from other tasks to the current repair. Any shifts in resource allocation are determined based on the user's assigned priority of events. For example, a medical evacuation sortie may be programmed to take precedence over an aircraft wash event.

In summary, the user provides a level of resources—manpower, equipment, facilities, and spares—that will be allocated to generate sorties for a known flight schedule. LCOM simulates the repair actions needed to produce the sorties from an aircraft pool. The LCOM analyst's task is to determine the quantity of resources that will meet mission requirements to a satisfactory level of service.[1]

---

[1] For further high-level background on LCOM's functions and processes, the reader is referred to the introductory material maintained by the 2nd Maintenance Requirements Squadron (2MRS) at Headquarters Air Combat Command, *LCOM Explained*, Langley AFB, Va.: HQ/ACC, 2MRS, December 1998.

Running the C-130 LCOM Model

The RAND team obtained LCOM maintenance network models for the C-130 from the 3rd Maintenance Requirements Squadron (3MRS) at AMC/HQ.[2] 3MRS originally developed these models to analyze manpower requirements for C-130 maintenance squadrons. The models include flight schedules for the missions the C-130 commonly supports, such as channel, Special Assignment Airlift/Air Mission, and Joint Airborne/Air Transportability Training flights. For deployed flying, the LCOM model includes missions for cargo and passenger transport, as well as sorties for medical evacuation. With these models, the team was able to replicate the results found in the LCOM reports published by 3MRS, and the AMC results became baseline values for this study.

With these task networks and sortie schedules in place, an LCOM scenario is then run with an arbitrarily large number of personnel. This informs the analyst of the total flying hours and the sortie success rate in the absence of manpower constraints. Based on the team's earlier research, manpower levels are then constrained to ensure the successful completion of at least 90 percent of the sorties that could be generated in the manpower-unconstrained case.[3] It is important to point out that this is a target the analyst selects and can readily change to satisfy any additional scenario objectives.

As this report discusses, not all operational facilities generate sorties. To determine the size of maintenance shops for a RMF that conducts periodic maintenance, such as ISO inspections, the team elected to use 95 percent of the manpower-unconstrained ISO inspection level as a lower-bound constraint. Again, the user chooses this target and can readily shift it to accommodate scenario requirements.

After determining shop sizes that meet each of the system's constraints, the analyst needs scaled-up manning values to account for MAFs. LCOM assumes that each individual in a maintenance shop is 100 percent available in his work shift for each day of the week the shop is open. The user applies a MAF to account for the factors that limit an individual's availability in an actual maintenance shop. Factors that come into play here are indirect labor (e.g., record keeping and classroom training), holidays, and work-week limitations (e.g., 40 hours a week in a peacetime environment or 60 hours under wartime conditions). Readers are referred to the 3MRS report for specifics on MAF computations, where the peacetime MAF is shown to be 1.038 for a five-day workweek and 1.511 for shops open seven days a week, and the sustained wartime MAF is computed to be 1.461.

MAFs are linear multipliers for an LCOM shop size. For example, if LCOM determined that a wheel-and-tire shop should be staffed by five people in a sustained wartime scenario, the actual manning of the shop should be $5 \times 1.461 = 7.3$, which then rounds up to eight individuals to satisfy the maintenance workload and other system constraints.

C-130 Models Specific to Operational Constructs

As indicated earlier, the AMC C-130 LCOM model is broken into three separate submodels to account for distinct operational environments. This subsection describes the original intent of

---

[2] Headquarters Air Mobility Command, *C-130 Logistics Composite Model (LCOM) Final Report: Peacetime and Wartime*, Scott AFB, Ill.: HQ AMC/XPMRL, 3MRS, February 12, 2001.

[3] McGarvey, Carrillo, et al., 2009.

these submodels, as well as how our team adapted AMC's models to account for the regionalized maintenance constructs that are described in this report.

The Home-Station Model. Earlier in this appendix, we discussed the various flying missions that can drive maintenance at home stations during peacetime. Additionally, the LCOM model addresses the periodic maintenance functions that can occur at the home station, such as aircraft wash, ISO inspections, HSCs, and refurbishment. Each of these activities, according to the 3MRS report, occurs on a regular schedule. In the peacetime environment of the home station, each aircraft receives a wash and corrosion check every 90 days, an HSC every 225 days, an ISO inspection every 450 days, and any refurbishment every 36 months.[4]

HSC Workload at the Home Station. As discussed in this report, we elected to adjust the scheduled maintenance activities that take place at the home station. The ISO inspection workload was moved to the RMF, as were the HSC and aircraft wash that align with each ISO inspection. Refurbishments were also removed from the home station and placed at the RMF, but their schedule was accelerated to align with every other ISO inspection (i.e., every 900 days, rather than every 1,040 days). While this accelerated schedule may overestimate the actual workload for refurbishment activities, it provides a conservatively large value for the manpower required at a centralized facility.

Each of these activities was selected for removal to an RMF because none of them directly affects sortie generation. Their consolidation at a regional facility offers the benefit and savings of personnel economies of scale, and the time spent in periodic maintenance can decrease, as discussed later in this appendix.

It is equally important to discuss why the HSCs were not fully consolidated at the RMF from the home station. As pointed out earlier, LCOM determined that HSCs consumed roughly 1 percent of the total MMH at the home station. This figure may seem to be small, but in light of evidence the research team acquired in visiting operational units, it seems roughly correct. In talking with MXGs, the team learned that it typically takes about five maintainers working for roughly five days to complete an HSC. While this may become a consequential workload during an actual HSC, what fraction of the maintenance squadron's total workload would we expect HSC to occupy?

Consider a notional, 8 PAA C-130 squadron, which would contain roughly 200 maintenance personnel. A squadron of this size requires about one HSC per month, so 5 divided by 200 maintainers conduct an HSC for about one-quarter of that month. With this rough "back of the envelope" calculation, HSCs would be estimated to absorb $(5 \div 200) \times (1 \div 4) = 0.6$ percent of the squadron's total MMH. Note that this value corresponds with LCOM's predicted value for the HSC workload. With this relatively small level of overall maintenance effort, it should be evident that a full consolidation of the HSC workload would achieve little benefit from any potential economies of scale found at an RMF.

Backshop Workload at the Home Station. This report has also pointed out that the off-equipment workload at the home station was a candidate for transfer to the RMF. Five key backshops were considered here: communication and navigation, electrical and environmental, guidance and control, pneudraulics, and propulsion. These shops were selected on the basis of the relatively low utilization rates in LCOM models of small- to medium-size squadrons (i.e.,

---

[4]   When the 3MRS report was originally released, the inspection interval for each HSC was 180 days and 365 for each ISO inspection. As of 2008, the inspection intervals are 225 and 450 days, respectively.

4 to 16 PAA). This made them ideal candidates for saving manpower through the economies of scale that consolidating maintenance at an RMF would provide.

Equally important, however, is that this specific potential consolidation we evaluated involved only the backshop component. The flightline maintenance component of the workload would remain on site. The relevant functions, skills, and expertise would thus remain available to the squadron.

An alternative to off-site consolidation of these shops is to merge backshop and flightline personnel at the home station. The notional construct here is a central squadron dispatch that would route personnel for each of the target shops. The primary maintenance function would be at the flightline to aid sortie generation. However, on observing that personnel from the shop are not needed on the flightline, the dispatch organization could route them to component repair activities. LCOM runs suggest these shops would provide significant savings for smaller squadrons (i.e., 4 to 8 PAA). LCOM indicates that, in some cases, the flightline crew could absorb the entire workload of a separate backshop without detracting from the primary goal of sortie generation.

However, as average flightline utilization increases for larger squadrons (greater than or equal to 12 PAA), it becomes less efficient to combine busy backshops with equally busy flightline crews. Thus, the possible benefits of the central dispatch model diminish, and the savings obtainable from centralization of backshops at an RMF become more attractive. To maximize the benefits of economies of scale, we thus opted to focus on the complete centralization of these five candidate backshops from the squadrons at the RMF.

To calculate the workload to be transferred from the home-station backshops to the RMF, the home-station model ran with increasing numbers of aircraft in conjunction with a commensurate increase in sorties demanded. In this way, backshop manpower could be computed in LCOM with the response to the increase in the component repair workload. To simplify the backshops' incorporation at the RMF, the backshops originating from the home stations were treated as separate from the shops dedicated to periodic maintenance activities, such as ISO inspections and refurbishments. This may not take full advantage of all potential personnel economies of scale, but the procedure generates a conservative estimate of those that are computed.

The FOL Model. As discussed previously, AMC's LCOM model for the FOL takes into account sortie activity typically found in theater. The FOL also conducts a regular HSC inspection at the standard 225-day interval. Given the calculations discussed above, consolidating the HSC workload at an RMF did not seem to offer any significant advantages.

However, as discussed in the 3MRS report, the FOL does not handle nongrounding equipment failures that occur in theater; the RMF does. We used the LCOM model to determine the MMH expected at the FOL, then converted this off-equipment workload on a shop-by-shop basis into an equivalent number of maintainers. As stated in the 3MRS report, the calculation takes into account such factors as the number of working hours per month, the maximum direct labor utilization rate, and the MAF, which was discussed earlier.

The number of maintenance personnel computed by this method is then added in a direct, linear fashion to the personnel computed from the backshops at the home station. The shared maintenance functions of the off-equipment workload seem to make this reasonable. While a simple linear addition of personnel may fail to capture possible economies of scale, especially at smaller RMFs, it staffs manpower conservatively at the RMF.

The RMF Model. AMC's original RMF construct envisioned that the central facility would handle only ISO inspections. To determine staff levels for the facility, we ran LCOM with a fixed number of ISO inspections to be completed in each simulated year. As noted earlier, the primary goal of the model runs is to ensure the successful completion of the requisite ISO inspection count without exceeding a maximum utilization rate in any of the individual maintenance shops.

One goal not explicitly constrained in running the RMF model is the average time required to complete each ISO inspection, which can also be thought of as an aircraft's average flow time in ISO inspection. While an analyst running the RMF model does not need to regulate ISO inspection flow times, LCOM does indeed report this metric. ISO inspection flow times are shown for various sizes of RMF in Figure C.1.

As shown in Figure C.1, ISO inspection flow times drop monotonically with an increase in the capacity of the repair facility. Average ISO inspection flow times drawn from REMIS for AFSOC and Little Rock AFB are plotted on the graph. These values are comparable to LCOM's predictions.

At first blush, the drop in flow time may seem unexpected. Because each ISO inspection needs a relatively constant number of MMH, it might be tempting to think that the hours required to complete an ISO inspection should also be constant. The LCOM analyst's goal, however, is to minimize the manpower required to accomplish the target workload at the RMF. A facility handling only a few ISO inspections each year would necessitate a relatively small requirement. Such a small cohort of manpower is likely to justify only one eight-hour work shift per day. As the workload increased with RMF capacity, more mechanics would be required, and it would thus become more reasonable to open multiple shifts a day for the shops. In the workload extremes, the RMF would range from a low utilization, eight-hour-per-day operation to a factory-style environment that is open 24 hours a day, seven days a week. Con-

Figure C.1
LCOM-Generated ISO Inspection Flow Times at an ISO Inspection–Only RMF

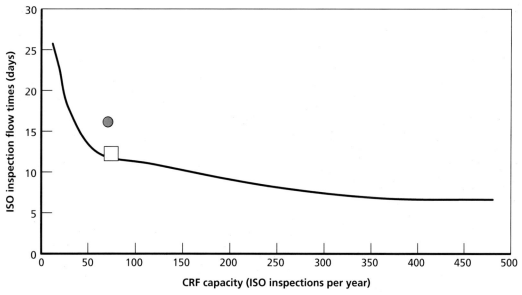

NOTE: The square indicates AFSOC; the dot, Little Rock.

sequently, the constant man-hours per ISO inspection that must stretch out over three weeks in a small facility can be compressed into less than a single week in a larger RMF. Indeed, this roughly corresponds to the 3-to-1 ratio moving from an eight-hour-a-day operation to a full three-shift, 24-hour work day.

It is important to note that, in the RAND construct, the RMF would also handle aircraft refurbishment, as well as the aircraft wash and corrosion checks that align with ISO inspection activities. To incorporate these features, we added the task networks corresponding with wash and refurbishment to the AMC ISO inspection model and adjusted shop staff levels until the number of maintenance activities appropriate to the size of RMF being studied was achieved.

Integer Linear Programming Model

To address the facility location problem, we formulated an integer linear program with an objective function that minimizes the annualized costs of the CRF network. The primary cost drivers for the problem are construction costs for aircraft hangars and equipment, transportation costs for shuttling aircraft between operating locations and CRFs, and maintenance personnel costs. The model's decision variables are the number, location, and hangar capacity of opened CRFs; the assignments of operating locations' aircraft to CRFs; and the manpower requirement at the CRF to perform the CRF workload. We assumed that each operating location is assigned to only one CRF.

We began by defining the sets used in the decision model. Let J be the full set of candidate CRF locations in the network, with index $j \in J$. Let I be the set of aircraft operating locations, with index $i \in I$. Let k be the set of capacity increments (personnel, facilities), with index $k \in K$, where k = 1,2,..., |K|. We assumed that repair capacity exists in the form of repair teams. As we discussed in Chapter Three, our definition of a repair team differs from our previous analyses. For the F-16, we fixed the average flow time for a phase and determined the manpower requirements necessary to maintain the flow time for a corresponding number of phases per year. However, given a desired phase flow time, it is difficult for CRFs supporting fewer PAA to maintain a fixed phase flow time without having a large number of work centers open for multiple shifts. The average productivity of these work centers is low relative to the work centers at large CRFs. Subsequently, we observed that increasing the size of CRFs offered substantial economies of scale in the maintenance manpower.

For the C-130 analysis, we approached the problem differently. Instead of fixing the flow time of the ISO inspection process and staffing the work centers accordingly, we chose to staff the work centers and the shifts with the minimum requirement such that no workers exceeded 60 percent utilization on average. Additional shifts are opened in the work centers as the size of operations increases. The *repair team* is defined as the number of additional personnel required to support an incremental increase in the number of ISO inspections supported at the CRF. In this case, as additional shifts are staffed at CRF work centers, the average ISO inspection flow rate increases incrementally. Therefore, we assumed that flow time is deterministic and a function of the number of repair teams at a given facility. Our assumption will influence the number of facilities required to support the CRF operations. As CRFs increase in size, and as work centers are staffed with multiple shifts, the number of ISO inspections per year per facility increases incrementally.

The demand at operating locations for ISO inspections and ISO inspection–related work depends on the TAI at each location and is linked to a 450-calendar-day cycle for ISO inspections. Therefore, we assumed that the demand is deterministic. Let λ_i equal the annual demand rate at operating location i for ISO inspection–maintenance. Let δ_{ji} equal the straight-line distance between operating locations j and CRF location i. Let ρ equal the total number of ISO inspections completed per repair team per year. Let

$$C^T_{ji}$$

equal the two-way transportation cost of sending an aircraft from operating location i to CRF j, and let

$$C^F_{jk}$$

represent the annual amortized facilities cost associated with incremental capacity value k at CRF j, including hangar space, shops, and equipment. Let

$$C^P_{ji}$$

represent the annual incremental cost of teams of personnel associated with capacity k at centralized phase facility j.

The problem has three sets of decision variables. First, let X_{jk} be the Boolean decision to open a CRF at location j with capacity k. As we previously discussed, the number of facility increments at a CRF will be a function of the flow times of the CRF, which will be a function of the number of aircraft assigned to the CRF. Second, let Y_{ji} equal the Boolean decision to assign demand location i to CRF j. Third, let Z_{jk} equal the Boolean decision variable to assign location j incremental capacity value k. The objective function is to minimize the sum of annual shuttle costs to and from CRF facilities, annual facility costs associated with CRFs, and the annual CRF personnel costs. The problem formulation is as follows

$$\min : \sum_{j \in J} \sum_{i \in I} C^T_{ji} Y_{ji} \lambda_i + \sum_{j \in J} \sum_{k \in K} C^F_{jk} X_{jk} + \sum_{j \in J} \sum_{k \in K} C^P_{jk} Z_{jk}$$

S.T.

$$\sum_j Y_{ji} = 1, \ \forall i \in I \tag{D.1}$$

$$\rho \sum_k Z_{jk} \geq \sum_i \lambda_i Y_{ji}, \ \forall j \in J \tag{D.2}$$

$$\sum_k X_{jk} \geq \sum_k Z_{jk}, \ \forall j \in J \tag{D.3}$$

$$X_{jk} \geq X_{jk+1}, \ \forall j \in J, \ k = 1, 2, \ldots, |K| - 1 \tag{D.4}$$

$$Z_{jk} \geq Z_{jk+1}, \ \forall j \in J, \ k = 1, 2, \ldots, |K| - 1. \tag{D.5}$$

Constraint (D.1) ensures that each operating location is assigned to a single CRF. Constraint (D.2) requires that the total repair capacity at CRF j be greater than the demand allocated to j. Constraint (D.3) ensures that each CRF j has matching facility capacity for the number of repair teams located at j. Constraints (D.4) and (D.5) enforce the purchase of facilities and repair capacity in a piecewise-linear relationship.

Bibliography

Air Force Civil Engineer Support Agency, *Historical Air Force Construction Cost Handbook*, Tyndall AFB, Fla.: Directorate of Technical Support, February 2004.

Air Force Instruction 65-503, *US Air Force Cost and Planning Factors,* Annex 4-1, "BY 2008 Logistics Cost Factors," Washington, D.C.: Department of the Air Force, February 22, 2006.

Air Force Instruction 65-503, *US Air Force Cost and Planning Factors,* Annex 19-2, "FY 2008 Standard Composite Rates by Grade," Washington, D.C.: Department of the Air Force, April 2007.

Clay, Shenita L., *KC-135 Logistics Composite Model (LCOM) Final Report: Peacetime and Wartime—Peacetime Update*, Scott AFB, Ill.: HQ AMC/XPMMS, May 1, 1999.

Department of the Air Force, "Fiscal Year (FY) 2009 Budget Estimate: Military Personnel Appropriation," briefing, February 2008.

DoD—*See* U.S. Department of Defense.

Drew, John G., Kristin F. Lynch, James M. Masters, Robert S. Tripp, and Charles Robert Roll, Jr., *Options for Meeting the Maintenance Demands of Active Associate Flying Units*, Santa Monica, Calif.: RAND Corporation, MG-611-AF, 2008. As of February 5, 2010:
http://www.rand.org/pubs/monographs/MG611/

Duncan Hunter National Defense Authorization Act for Fiscal Year 2009, Section 324 of H.R. 5658, 110th Congress, 2008.

Government Services Administration, Awarded Contractors Under Domestic Delivery Services, website, October 26, 2006. As of February 17, 2010:
http://apps.fas.gsa.gov/services/express/awarded-cont.cfm

Headquarters Air Combat Command, *LCOM Explained*, Langley AFB, Va.: HQ/ACC, 2MRS, December 1998.

Headquarters Air Combat Command (HQ ACC), Director of Plans and Programs, Manpower and Organization Division (XPM), *Statement of F-16 Block 40 Aircraft Maintenance and Munitions Manpower*, Langley AFB, Va.: HQ ACC/XPM, August 2003.

Headquarters Air Mobility Command, *C-130 Logistics Composite Model (LCOM) Final Report: Peacetime and Wartime*, Scott AFB, Ill.: HQ AMC/XPMRL, February 12, 2001.

Hoehn, Andrew R., Adam Grissom, David A. Ochmanek, David A. Shlapak, and Alan J. Vick, *A New Division of Labor: Meeting America's Security Challenges Beyond Iraq*, Santa Monica, Calif.: RAND Corporation, MG-499-AF, 2007. As of February 5, 2010:
http://www.rand.org/pubs/monographs/MG499/

International Air Transport Associates (IATA), "Jet Fuel Price Monitor." As of July 30, 2008:
http://www.iata.org/whatwedo/economics/fuel_monitor/index.htm

Julian, Jon, email, AMC/A8PF, September 2008.

McGarvey, Ronald G., Manuel Carrillo, Douglas C. Cato, Jr., John G. Drew, Thomas Lang, Kristin F. Lynch, Amy L. Maletic, Hugh G. Massey, James M. Masters, Raymond A. Pyles, Ricardo Sanchez, Jerry M. Sollinger, Brent Thomas, Robert S. Tripp, and Ben D. Van Roo, *Analysis of the Air Force Logistics Enterprise: Evaluation of Global Repair Network Options for Supporting the F-16 and KC-135*, Santa Monica, Calif.: RAND Corporation, MG-872-AF, 2009. As of February 5, 2010:
http://www.rand.org/pubs/monographs/MG872/

McGarvey, Ronald G., James M. Masters, Louis Luangkesorn, Stephen Sheehy, John G. Drew, Robert Kerchner, Ben Van Roo, and Charles Robert Roll, Jr., *Supporting Air and Space Expeditionary Forces: Analysis of CONUS Centralized Intermediate Repair Facilities*, Santa Monica, Calif.: RAND Corporation, MG-418-AF, 2008. As of February 5, 2010:
http://www.rand.org/pubs/monographs/MG418/

National Defense Authorization Act for Fiscal Year 2008, Public Law 110-181, January 28, 2008. As of June 8, 2010:
http://www.govtrack.us/congress/bill.xpd?bill=h110-4986

Ronald W. Reagan National Defense Authorization Act for Fiscal Year 2005, Public Law 108-375, October 28, 2004. As of June 8, 2010:
http://www.govtrack.us/congress/bill.xpd?bill=h108-4200

Tripp, Robert S., Ronald G. McGarvey, Ben D. Van Roo, James M. Masters, and Jerry M. Sollinger, *A Repair Network Concept for Air Force Maintenance: Conclusions from Analysis of C-130, F-16, and KC-135 Fleets*, Santa Monica, Calif.: RAND Corporation, MG-919-AF, 2010. As of April 8, 2010:
http://www.rand.org/pubs/monographs/MG919/

U.S. Air Force, *Expeditionary Logistics for the 21st Century*, Washington, D.C., April 1, 2005.

U.S. Air Force, Deputy Chief of Staff for Installations and Logistics Directorate of Transformation (AF/A4I), *Logistics Enterprise Architecture (LogEa) Concept of Operations*, May 24, 2007.

U.S. Air Force, Manpower and Personnel Integration (AF/A1MZ), TPERDET META DATA.doc, November 2007.

U.S. Department of Defense, *Quadrennial Defense Review Report*, Washington, D.C., September 30, 2001.

———, *BRAC Commission Action Brief*, Washington, D.C., September 1, 2005.

———, Program Budget Decision 720 ("Air Force Transformation Flight Plan"), Washington, D.C., December 2005.

U.S. Department of Defense and the Joint Chiefs of Staff, *Mobility Capabilities Study*, Washington, D.C., December 19, 2005; not available to the general public.

U.S. House of Representatives Committee on Armed Services, Panel on Roles and Missions, *Initial Perspectives*, Washington, D.C., January 2008. As of December 11, 2008:
http://militarytimes.com/static/projects/pages/hasc_roles_missions0308.pdf

Wynne, Michael W., Secretary of the Air Force, "Air Force Smart Operations 21," Letter to Airmen, Washington, D.C. August 8, 2006. As of July 26, 2009:
https://acc.dau.mil/CommunityBrowser.aspx?id=140491

Wyoming National Guard Public Affairs, "30th Airlift Squadron settles in at Wyoming Air National Guard," Cheyenne: Wyoming Air National Guard 153rd Airlift Wing, December 4, 2006. As of December 11, 2008:
http://www.amc.af.mil/news/story.asp?id=123033749